THE

MOMENT

V
A
F
B

MORDECHAI FEINGOLD

THE

MOMENT

ISAAC NEWTON AND THE MAKING OF MODERN CULTURE

THE NEW YORK PUBLIC LIBRARY/OXFORD UNIVERSITY PRESS
NEW YORK OXFORD 2004

Published on the occasion of the exhibition

THE NEWTONIAN MOMENT:
SCIENCE AND THE MAKING
OF MODERN CULTURE

presented at The New York Public Library,
Humanities and Social Sciences Library
D. Samuel and Jeane H. Gottesman
Exhibition Hall
October 8, 2004–February 5, 2005

This exhibition has been organized by The New York Public Library in cooperation with Cambridge University Library.

Generous support for this exhibition has been provided by The Horace W. Goldsmith Foundation–Robert and Joyce Menschel; Robert and Mary Looker; Mr. and Mrs. Ira D. Wallach; and The Dibner Fund.

Support for The New York Public Library's Exhibitions Program has been provided by Pinewood Foundation and by Sue and Edgar Wachenheim III.

www.nypl.org

www.oup.com/us

Oxford University Press
Oxford New York
Auckland Bangkok Buenos Aires Cape Town
Chennai Dar es Salaam Delhi Hong Kong
Istanbul Karachi Kolkata Kuala Lumpur Madrid
Melbourne Mexico City Mumbai Nairobi
São Paulo Shanghai Taipei Tokyo Toronto

Published in 2004 by Oxford University Press, New York, in association with The New York Public Library

ISBN 0-19-517735-5 (cloth)
ISBN 0-19-517734-7 (paper)

Cataloging-in-Publication Data available from the Library of Congress

Printed in China on acid-free paper

Karen Van Westering *Manager, NYPL Publications*
Anne Skillion *Senior Editor*
Barbara Bergeron *Editor*
Kenneth Benson *Editor*
Jennifer Woolf *Photography Coordinator*

Interior designed by Marc Blaustein

To the memory of
Frank E. Manuel (1910–2003)
and I. Bernard Cohen (1914–2003),
mentors and fellow travelers.

CONTENTS

FROM THE PRESIDENT

Exhibitions on almost any scale are really vast exercises in cooperation, engaging the talents, enterprise, and good will of many individuals and organizations. The New York Public Library's exhibit on Sir Isaac Newton and this companion volume, *The Newtonian Moment: Isaac Newton and the Making of Modern Culture*, represent just such an effort. It is my pleasant task to thank many of those individuals who made all of it happen, and with such success.

The inspiration to mount an exhibit at our Library on Newton and the revolution in worldview that his work inspired came with the news, in 2000, that the Cambridge University Library had acquired the Macclesfield Collection of scientific papers and letters. Cambridge, Newton's university, already possessed stunning manuscripts and printed materials related to his genius and scholarship, and this latest important purchase inspired me to think about a large-scale exhibit in New York that would showcase some of these irreplaceable documents, many of which had been privately owned until then.

Peter Fox, who directs the Cambridge University Library with very great skill, responded most enthusiastically to my idea and facilitated our initial visit to Cambridge in the summer of 2001 to see an extraordinary sampling of Newton's manuscripts, as well as a marvelous exhibit on Newton at work that the University Library had mounted.

Once Peter Fox had agreed to lend these materials to The New York Public Library, our next task was to identify the ideal curator for our exhibit. I turned for advice to Anthony Grafton, the distinguished Princeton historian of European intellectual history, whose association with our Library has been a long and happy one. He suggested that I invite Mordechai Feingold, Professor of History at the California Institute of Technology.

Tony's suggestion was characteristically brilliant: when H. George Fletcher, Brooke Russell Astor Director of Special Collections and the administrator of our exhibitions program, and I met with Professor Feingold, we knew instantly that he had not only the intellectual energy needed to coordinate an exhibition as challenging as this one, but also the didactic skill necessary to bring Newton's astonishing advances in mathematics and science to life for the contemporary lay person. His accompanying book is an original exploration of the ways in which Newton's thought has permeated Western culture since the Enlightenment. I am deeply indebted to Professor Feingold for having produced both this wonderful book and what is surely one of the most significant and exciting exhibits at The New York Public Library in a decade.

I am also happy to thank all of our generous donors who made this exhibit possible. The New York Public Library continues to be committed to the principle of free access to all the information it contains. That noble precept extends to our exhibitions program as well; hence, in the absence of entrance fees, exhibits have to be underwritten by donors who place a high value on broad democratic access to knowledge. *The Newtonian Moment* therefore owes its existence to the generous support of The Horace W. Goldsmith Foundation–Robert and Joyce Menschel; Robert and Mary Looker; Mr. and Mrs. Ira D. Wallach; and The Dibner

Fund. Support for The New York Public Library's Exhibitions Program has been provided by Pinewood Foundation and by Sue and Edgar Wachenheim III.

The extraordinary variety of materials included in *The Newtonian Moment* is owing to the generosity of a substantial number of libraries, museums, and individuals. The Cambridge University Library deserves special thanks not only for lending exceptional treasures but also for covering all expenses relative to their travel to The New York Public Library. This exemplifies a remarkable level of interinstitutional, international cooperation and I am especially grateful to Peter Fox and the Cambridge University Trustees for this magnificent contribution to *The Newtonian Moment*.

I am also grateful to the following organizations and individuals for the loan of materials from their collections: Adler Planetarium & Astronomy Museum, Chicago; The Burndy Library, Dibner Institute for the History of Science and Technology, Cambridge, Massachusetts; Butler Library, Columbia University, New York; California Institute of Technology Archives, Pasadena; Cambridge University Library, England; The Colonial Williamsburg Foundation; Mr. and Mrs. Robert Gordon; Collection of Historical Scientific Instruments, Harvard University, Cambridge, Massachusetts; Houghton Library of the Harvard College Library, Cambridge, Massachusetts; Jewish Theological Seminary of America, New York; The Metropolitan Museum of Art, New York; The Pierpont Morgan Library, New York; The New-York Historical Society; Smithsonian Institution Libraries, Washington, D.C.; and a private collection.

Our thanks to the following institutions for permission to reproduce in the exhibition works of art from their collections: Alte Pinakothek, Munich; Bibliothèque nationale de France, Paris; Civici Musei d'Arte e Storia di Brescia; Fitzwilliam Museum, Cambridge; Galleria Doria Pamphilij, Rome; Niedersächsisches Landesmuseum, Hanover; Royal Society of Arts, London; and Tate Britain, London. Professor Feingold has mentioned in his own acknowledgments the contributions of additional institutions to the creation of this book.

Finally, I wish to express my thanks to the administrative, curatorial, exhibitions, publications, and design staffs of The New York Public Library for all their remarkable efforts on behalf of Newton.

PAUL LECLERC
President, The New York Public Library

Joseph Wright of Derby's "A Philosopher Giving a Lecture on the Orrery" (1766) illustrates the manner in which the scientific lecture-demonstration had become a popular form of public entertainment by the middle of the eighteenth century. Wright's philosopher bears a marked resemblance to Isaac Newton. – Derby Museum and Art Gallery, UK/Bridgeman Art Library

INTRODUCTION

In 1787, on the centenary of the first edition of Isaac Newton's *Principia*, the German philosopher Immanuel Kant published the second edition of his celebrated *Critique of Pure Reason*, with an extensive new introduction. His intent to turn metaphysics into a "science," Kant announced, involved altering "the procedure which ha[d] hitherto prevailed in metaphysics, by completely revolutionizing it in accordance with the example set by the geometers and physicists." The revolution Kant had in mind was modeled on the revolution that Newton had introduced into the natural sciences. Like Newton, Kant considered the natural sciences to be strictly "founded on *empirical* principles." So, too, his conception of the methodology and procedures proper to the practice of the natural sciences echoed Newton's commitment to experimentalism guided by reason as the means to establish the laws of nature.[1]

The foremost thinker of the eighteenth century, Immanuel Kant (1724–1804) sought to transform philosophy by following the example set by mathematicians and physicists. Kant would model his revolution on the revolution that Newton had introduced into the natural sciences. – NYPL–Print Collection

Yet nowhere in the introduction did Kant mention Newton by name, except in one footnote. Quite simply, there was no need for him to do so. By 1787, the conception of the natural sciences laid out by Kant would have been instantly identifiable to contemporaries as "Newtonian science." Nor would these contemporaries have failed to detect in Kant's determination to make metaphysics a "science" yet another attempt to extrapolate the Newtonian success story to other domains. This subtext of Kant's introduction, a hundred years after the *Principia*, is testimony to the enduring legacy of Newton's spectacular contributions to mathematics and natural philosophy. The discovery of the calculus, the articulation of a radical new theory of light and colors, and the unification of terrestrial and celestial mechanics under a single law had set the natural sciences firmly on a new course and, even more dramatically in terms of the human story, lifted a professor of mathematics to unprecedented heights of celebrity. Had he lived in antiquity, contemporaries had little doubt, Newton would surely have been deified.

Clearly, then, Newton's influence transcended the domain of science. During a time when the mathematical sciences and natural philosophy were integral to a much broader encyclopedia of knowledge, the apparent success of these domains set an example of so-called superior knowledge for other disciplines to emulate: the search for rational, universal principles became the *modus vivendi* for all researchers, regardless of field. Naturally, some dissented from this summons to reorient knowledge, sparking heated debates over the applicability of mathematics (and physics) to other areas of science, as well as between the sciences and the humanities over the kind of knowledge most worth having. Notwithstanding these burgeoning controversies, or perhaps because of them, for friends and foes alike Newton became an icon to be emulated or rejected, revered or excoriated – but always there to contend with. Hence, the era of Enlightenment and Revolution may be viewed as the Newtonian Moment, understood as denoting the epoch and the manner in which Newtonian thought came to permeate European culture in all its forms.[2]

This volume attempts to narrate the conception and diffusion of Newton's ideas, and the tensions and often public clashes they have engendered. Conceived as a companion volume to an

Attempts to go beyond Newton and to provide more detailed accounts of the universe had become fashionable by the mid-eighteenth century. Among the first was Thomas Wright's *An Original Theory or New Hypothesis of the Universe* (London, 1750), in which the author announced his intention of solving the problem of the Milky Way, thereby offering "a regular and rational Theory of the known Universe." Plate XXXI was intended to visualize "a partial View of [the] immensity" of a universe full of solar systems and planetary worlds. - NYPL–Spencer Collection

exhibition at The New York Public Library, the volume necessarily offers but a sampling of some of the many facets that constituted the Newtonian "moment."

Contrary to the common perception, Newton was not the "solitary and dejected" autodidact he is commonly perceived to have been. Nor was Cambridge University, where Newton lived for thirty-five years (1661–96), the bastion of scholasticism and intellectual stagnation it is often characterized as. In fact, Cambridge contributed significantly to the maturation of Newton's genius. The university's well-rounded and humanistically informed curriculum proved indispensable to Newton's grounding in the culture of erudition, and propitious to the formation of his scientific methodology and distinct style of reasoning. Cambridge also provided Newton with access to books and like-minded colleagues – above all, his mentor, friend, and patron Isaac Barrow. In this sense Newton truly stood "on the shoulders of giants," as he once wrote (albeit tongue-in-cheek) to Robert Hooke. Much of Newton's genius consisted of his remarkable ability to simultaneously consume and transform any knowledge he acquired. Consequently, his celebrated *anni mirabiles* (wondrous years) back in Lincolnshire during the plague (1665–66) were not cut off from his Cambridge experience, but were its natural extension. Samuel Johnson, therefore, was surely correct to conclude that Newton stood alone "merely because he had left the rest of mankind behind him, not because he deviated from the beaten track."

The unveiling of Newton's sensational miniature reflecting telescope before the Royal Society of London in the closing days of 1671 catapulted Newton to European fame. Gratified by the enthusiastic reception of his "toy," Newton agreed to publish his revolutionary theory of light and colors. The ensuing controversies over the verity of the theory, however, made Newton vow never to appear in print again. Only owing to the considerable scientific and diplomatic skills of Edmond Halley did Newton agree to write, and then publish, the *Principia* (1687). Seventeen more years elapsed before the *Opticks* (1704) finally appeared. Both works generated as much excitement as controversy. Subsequent editions clarified and elaborated on certain Newtonian concepts, as well as responded to criticisms. The eruption of the calculus priority dispute between Newton and Gottfried Wilhelm Leibniz complicated the response to the two works, prejudicing Leibniz's disciples against the central tenets of Newton's masterpieces, and causing a rift between English and Continental mathematicians. Ironically, those very disciples proceeded to translate the *Principia* into the Leibnizian form of the calculus (differential equations), thus creating the necessary mathematical tools for the future assimilation and advancement of Newton's ideas. Newton's English disciples, for their part, began rendering the *Opticks* and, especially, the *Principia* into the more accessible format of commentaries, aimed at those with only a modicum of mathematical background.

In his old age, Newton was known to boast that he had made the *Principia* purposely difficult in order to stave off "smatterers" in mathematics. He need not have tried. The incomprehensibility of the treatise, however, derived more from the theory and structure of the book than from the need to master a new language of mathematics or to assimilate the mystifying concept of action at a distance. Newton's refusal to offer a mechanical cause to account for universal gravitation, or to provide an underlying metaphysical framework, further unsettled contemporaries. Accustomed to thinking about natural philosophy in terms of causes and a priori reasoning, they bristled at Newton's suggestion that certain knowledge could be derived directly from the phenomena of nature, and that there was no need to "feign hypotheses." A lengthy process of assimilation, therefore, was necessary before conversion to Newtonianism was possible, especially as chauvinistic overtones compounded these inherent difficulties of comprehension: acceptance or rejection of Newtonian ideas was as likely to be made along nationalistic lines as on the merits of the case by German proponents of Leibniz, or by Frenchmen who balked at the spectacle of the dethroning of Descartes by an Englishman.

The daunting effect of the underlying mathematics on the diffusion of Newtonian ideas was alleviated by the inventiveness of English and Dutch scientific practitioners in designing scientific instruments – and devising ingenious experiments – capable of establishing Newtonian principles. "Forces" and the laws of motion suddenly became every bit as visual (and demonstrable) as the refrangibility of white light through a prism. Thanks to

In this 1720 broadside, *A Scheme of the Solar System with the Orbits of the Planets and Comets*, the English natural philosopher (and anti-Trinitarian heretic) William Whiston grafted the purported orbits of several comets onto a Newtonian scheme of the solar system. – Courtesy of Adler Planetarium & Astronomy Museum, Chicago, Illinois

the efforts of these university professors, instrument makers, and itinerant lecturers, the scientific lecture-demonstration became the backbone of university instruction in the natural sciences as well as a fashionable form of public entertainment among aristocrats and members of the middle class. The popularity of science increased even more by the middle of the eighteenth century, with the harnessing of electricity and the coming into vogue of natural history.

By the mid-eighteenth century, a powerful "image" of Newton had come to inform the Enlightenment and inspire generations of philosophers and men of letters as well as mathematicians and natural philosophers. For the English empiricists; for the Scottish Common Sense philosophers; for the French *philosophes*, and for the various members of the so-called counter-Enlightenment, Newton was first and foremost an "emblem" of a new era. Whether they admired and sought to imitate the great Englishman or disagreed with and attacked him, the actual encounter was guided by their perception that he had provided a new point of departure for all future probes into the three major domains of human inquiry: man, nature, and God.

Emulation naturally led to apotheosis. Edmond Halley initiated the process in a poem he contributed to the first edition of the *Principia*, the final line of which decrees: "no closer to the gods can any mortal rise." For the next 150 years, admiration of Newton bordered on idolization; he was immortalized in verse, carved in stone, his bust prominent in the "temples of worthies" that proliferated in aristocratic gardens of the eighteenth century. It was a common practice to hang the portrait of the great Englishman in the study of a scientist or a man of letters, thereby paying homage – and perhaps hoping for inspiration. Another way to indicate an intellectual link was to include a bust of Newton in a commissioned portrait, or at the very least inscribe his name on the spine of a book depicted in the background. Not a few artists sanctified Newton's genius or his contribution to science in their paintings, while shrewd entrepreneurs appropriated his name or portrait to adorn their firm's logo. With time, the historical Newton receded into the background, overshadowed by the very legacy he helped create. Newton thus metamorphosed into science personified.

William Blake's highly ambiguous stance toward Newton – whom he recognized to be a towering genius even as he excoriated the influence of his pernicious "single vision" on both religion and literature – is vividly captured in Blake's "Newton," from 1795. A very handsome (though contorted) Newton is depicted seated on a rock, underneath the "sea of time and space" representing materialism, busy drawing geometrical figures with his compasses. And yet, the loathsome abstract designs are written on a scroll, which signifies creativity. Here and elsewhere for Blake, Newton is the misguided genius whose mechanical universe left no room for the imagination or for God, but who would ultimately find his prominent place in heaven.

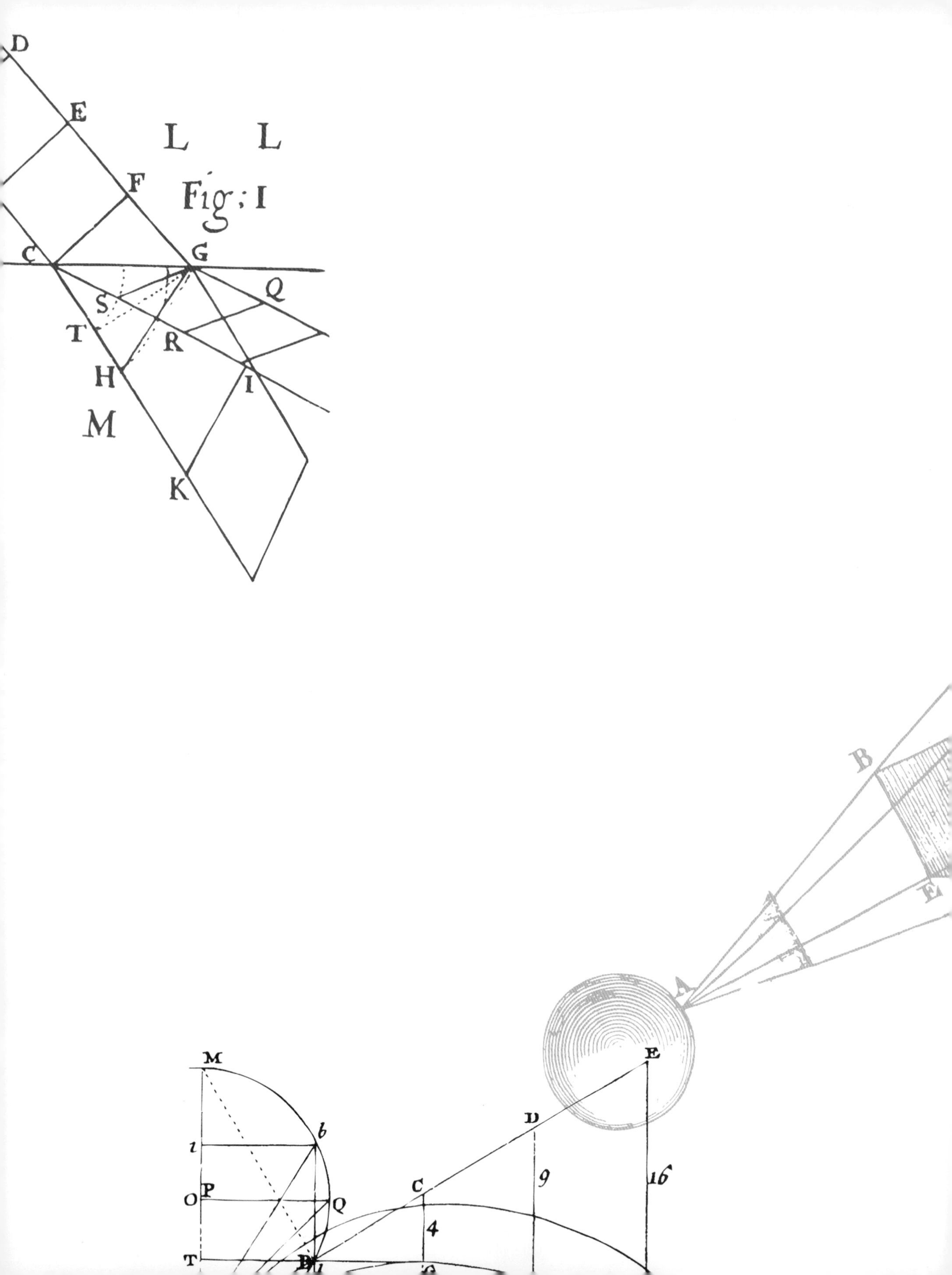
D
E
L
L
F
Fig; I
C
G
Q
S
T
R
H
I
M
K
B
E
A
M
E
b
D
1
9
16
C
O
P
Q
4
T
B

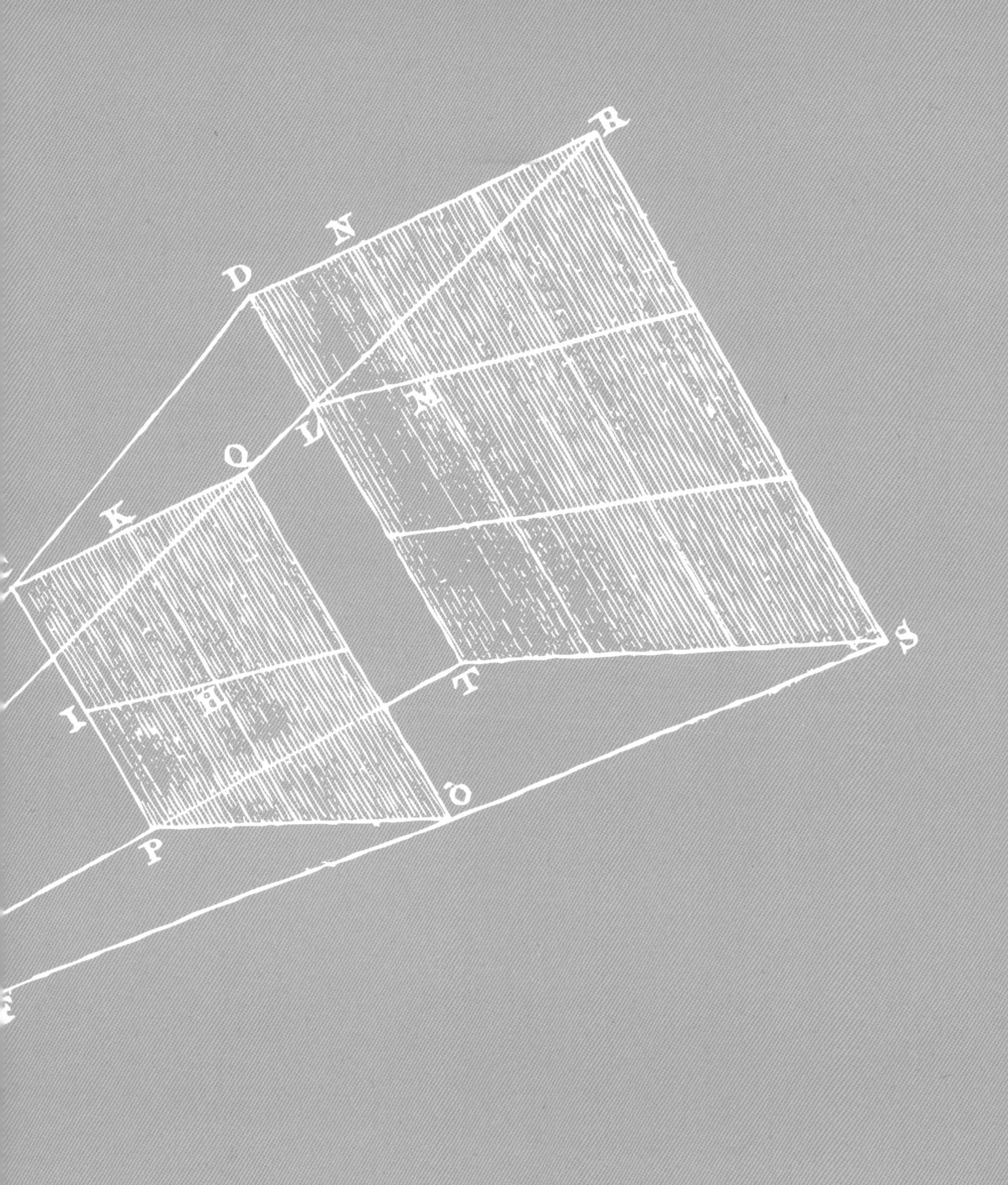
R
N
D
L
N
Q
K
I
H
T
S
O
P

1

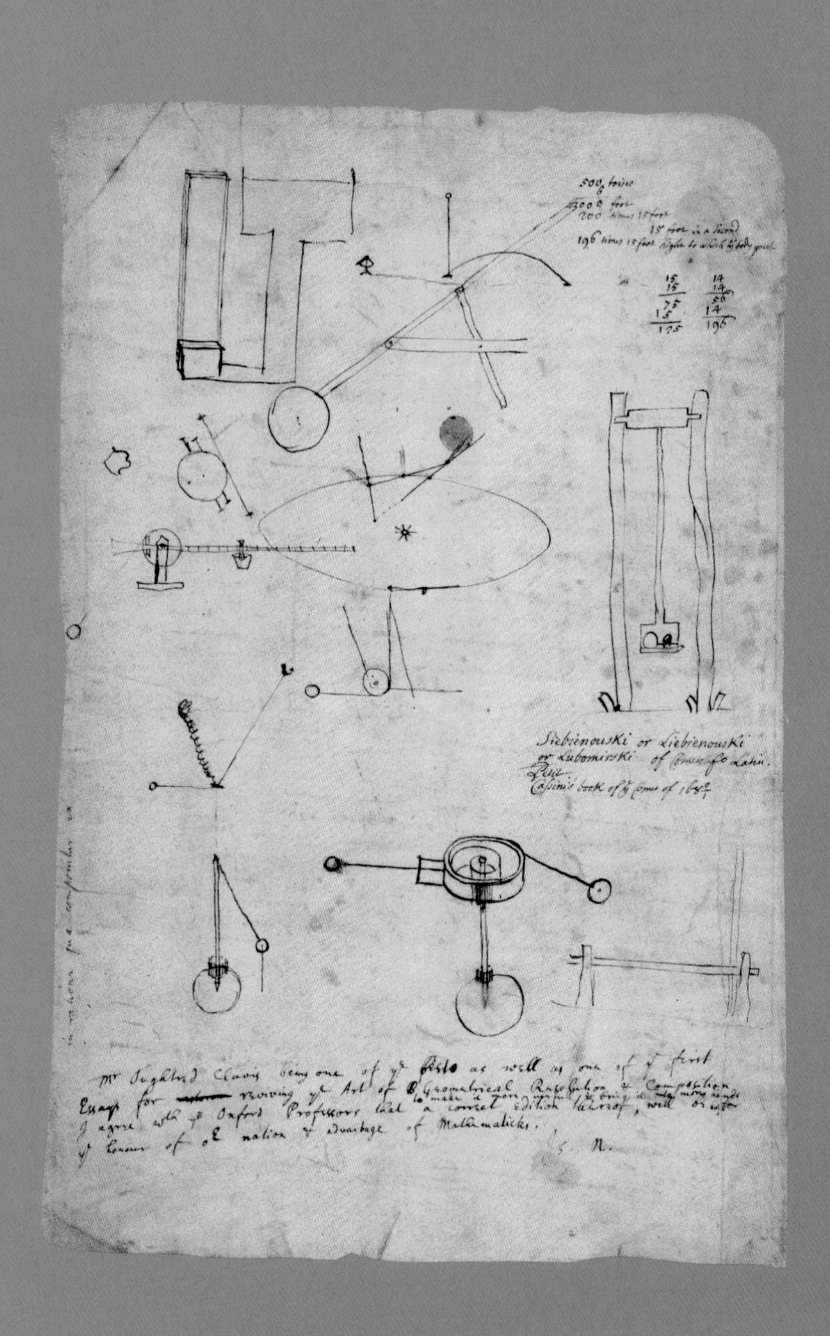

These sketches of an apparatus, drawn by Newton, are among the *Principia* papers (Cambridge, 1680s). – Cambridge University Library, MS ADD 3965, fol. 35v

1

THE APPRENTICESHIP OF GENIUS

Isaac Newton took three days to traverse the roughly sixty miles separating his family's manor house in Woolsthorpe, Lincolnshire, and Cambridge University. He arrived there on the evening of June 4, 1661, and presented himself the following day to the dean of Trinity College. After entering his name in the College's admissions book, Newton was assigned a tutor – Benjamin Pulleyn, a respected classicist who would become Regius Professor of Greek in 1674 – and directed to his chamber. Born in the early hours of Christmas Day 1642, Newton was eighteen-and-a-half years old that summer, somewhat older than most incoming undergraduates – a reflection of his twice-widowed mother's reluctance to make her son a scholar instead of a helpmate in managing the family estate. Indeed, Hannah Newton pulled her son out of grammar school early; she eventually relented, owing largely to the persistence of his schoolmaster, Henry Stokes, who recognized his charge's budding genius and "never ceased remonstrating to his mother what a loss it was to the world, as well as a vain attempt, to bury so extraordinary a talent in rustic business." The youth returned to Grantham School, where he lodged with Stokes and prepared himself for higher education.

The intellectual and psychological distance between the world of Newton's youth and the world of Cambridge was immeasurably greater than any geographical distance. The rusticity and provincialism of the Lincolnshire countryside contrasted sharply with the cosmopolitanism and intellectual sophistication that quickly overtook Cambridge after 1660, when nearly two decades of Puritan austerity and religious zeal came to an end with the restoration of the Stuart monarchy. Trinity College, in particular, by virtue of its size, its high proportion of upper-class students, and the intellectual stature of its members, past and present, helped set the university's academic and social tone. Traditionally, it has been taken for granted that Newton was ill-suited – intellectually as well as temperamentally – for such an urbane environment. To make such an assumption, however, is to misunderstand the nature of the early modern university and to miscon-

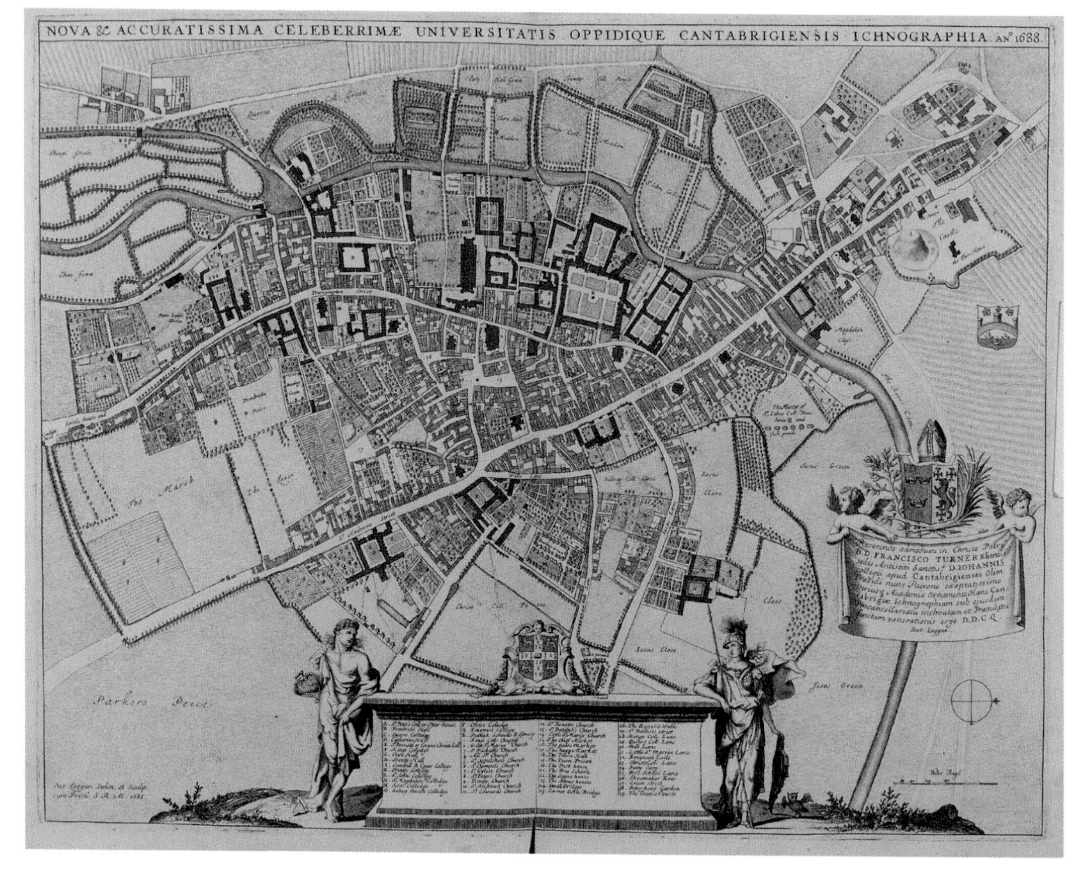

Trinity College, where Newton enrolled in 1661, is shown at center right on this map of Cambridge University, from *Cantabrigia illustrata* by David Loggan (Cambridge, 1690). – NYPL–Print Collection

strue the few anecdotes that have survived from that period of Newton's life. For example, Newton was admitted to Trinity as a subsizar, the lowliest of college ranks, usually reserved for poor students expected to serve their social betters in return for tuition and board. That Newton entered college under such conditions, however, was not because his mother was impecunious; she was in fact fairly prosperous. She was simply loath to throw away good money on status, especially since Newton was expected to be attached to the mostly absentee fellow (and family friend) Humphrey Babington and thus be released from the onus of menial tasks associated with the subsizar status. Certainly, the surviving records make clear that Newton's finances were healthy enough to allow him not only modest indulgence in conspicuous consumption, but also the resources to become an enterprising moneylender among Trinity undergraduates. Newton's motivation in such activity, it might be added, was not necessarily monetary profit; he seems never to have charged interest. The enticement was the means it offered to position himself among – or above – the more affluent undergraduates.

Just as Newton never assumed the life of servitude often attributed to him, so, too, his common depiction as an outcast ignores a more nuanced reality. Newton may not have been a socialite, and he certainly could immerse himself in hard work virtually to the point of abandon; but neither was he a recluse. An anecdote detailing how he and John Wickins became chamber fellows in mid-1663 is a case in point. As the latter's son recounted some sixty years later, Wickins had been assigned a "very disagreeable" roommate; one day, walking in the college garden in order to avoid him, he encountered Newton, "solitary and dejected" for precisely the same reason. The two then "agreed to shake off their present disorderly Companions and Chum together." Such an account of an experience familiar to countless undergraduates past and present hardly warrants the portrayal of Newton as a student in a perpetual state of "isolation and alienation." Equally unwarranted is the extrapolation from the recollections of an amanuensis – that Newton "always kept close to his studys," rarely received visitors, never took "any Recreation or Pastime," often forgot his meals, and was careless in his attire – that such patterns characterized Newton's thirty-five-year career at Cambridge. The amanuensis, Humphrey Newton (no relation), served his namesake from 1685 to 1690, when Isaac was involved first in writing the *Principia* and, immediately thereafter, in leading the university in opposing the Catholic policies of James II. The mental and emotional strain of composing his masterpiece in less than two years, and then turning to a holy war against his monarch, is hardly the proper yardstick against which to determine Newton's sociability.[1]

A perception that Newton was not a recluse and that his relations with peers and teachers were not necessarily different in kind from those of other students is crucial if we are to make an equitable assessment of Cambridge's contribution to his genius. For most commentators, such a contribution has been viewed, if at all, in negative terms: Newton the undergraduate embraced alienation to flower

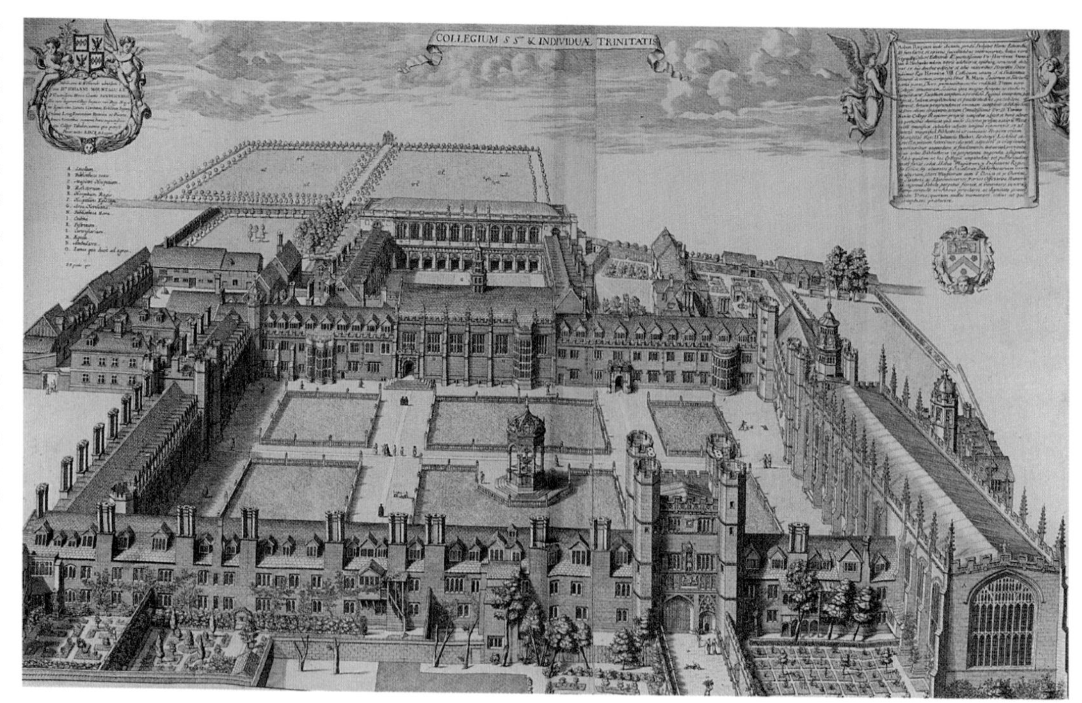

The cosmopolitanism and intellectual sophistication of Trinity College, shown here, contrasted sharply with the world of Newton's youth. Newton's room was on the first floor of staircase E, to the right of the great gate. From *Cantabrigia illustrata* by David Loggan (Cambridge, 1690). – NYPL–Print Collection

in the allegedly inhospitable academic environment in which he found himself. Newton faced his share of ordeals, to be sure, including an acute religious crisis suffered on Whitsunday 1662, which incited him to draw up a list of sins he believed he had committed before and after coming to Cambridge. Of the former, a few recalled such dark thoughts of yore as when he threatened "to burne" his stepfather and mother "and the house over them," or when he felt himself "Wishing death and hoping it to some." His professed Cambridge transgressions, in contrast, attest to his leading a relatively normal and worldly life. He agonized that this lifestyle was at odds with the stern religiosity in which he was reared: swimming on the Sabbath; baking pies on Sunday night; engaging in "Idle discourse on Thy day and at other times"; squirting water (presumably on fellow students) "on Thy day"; being inattentive at sermons; and setting his heart on money, learning, and pleasure "more than [on] thee" – in short, the very failings that bear out a moderately successful social acculturation.[2] As with other undergraduates who experienced such crises, Newton seems to have assuaged his conscience rather quickly and resumed his integration into Trinity undergraduate life.

The relative normalcy of Newton's social life applies to his scholarly life as well. By the time Newton came up to Cambridge, the curriculum had been radically transformed from the course in fashion a century earlier. Though for pedagogical convenience the English universities clung to scholasticism in form, the content was restructured. The entire arts and sciences curriculum was now compressed into the undergraduate course, thereby transforming the M.A. sequel into a course of independent study, aimed at expanding and elaborating the foundations of knowledge previously acquired. As a result, the character of the curriculum became quintessentially humanistic, with a pronounced focus on the languages and literatures of ancient Greece and Rome. Such an orientation dovetailed with the sustained effort to provide students with adequate grounding in the full encyclopedia of knowledge – that is, to acquaint them with the terms of art in the various disciplines and with a modicum of their respective subject matters. The ideal informing the undergraduate curriculum, then, was that of the erudite "general scholar," who could later acquire whatever specialized knowledge he wished. Newton became a beneficiary of this ideal, as his successful applications in later years to historical, chronological, and theological studies – leaving aside his mathematics and natural philosophy – attest.[3]

Irrespective of social background or the length of university stay, then, all undergraduates pursued a virtually identical course of studies. Mornings over several months were allocated, variously, to logic, natural and moral philosophy, history and chronology, geography and mathematics. The afternoons, in contrast, spanning the student's entire undergraduate career, were reserved for immersion in language and literature – testimony to what truly animated early-modern educated men. The tutor was assisted in his duties by several college lecturers who offered elementary instruction in the staples of undergraduate education – logic, rhetoric, natural philosophy, and mathematics – and by university professors who offered more advanced instruction for those seeking it. To a large extent, however, the acquisition of learning was accomplished by private study, even on the undergraduate level. The tutor's duty was more to guide and to supervise than to teach. Having ascertained that the basic principles of the disciplines were grasped by the student, the tutor assumed the role of director of studies, overseeing the more or less independent consolidation of higher-level skills. This gradually attained self-sufficiency became the norm after graduation, when the newly created Bachelor of Arts was relieved of both tutor and daily supervision. In the vivid language of a contemporary observer, "then they are turned loose, and with their paper-barks committed to the great ocean of learning."[4]

Such a system allowed students to benefit from the instruction of fellows other than their official tutors. This was especially significant for such specialized topics as Semitic languages and the mathematical sciences, when the official tutor could opt to "farm out" his charges to colleagues who were better qualified to teach them. The existence of such a flexible structure helps explain the nature of the early relationship between Newton and Isaac Barrow, who had been Regius Professor of Greek (and Gresham Professor of Geometry) prior to his election in late 1663 as first Lucasian Professor of Mathematics. Relying on information gleaned from Newton himself, his friend and early biographer William Stukeley singled out Barrow as Newton's tutor, observing that if Newton "did not take a byass in favor of mathematical studys from him, at least he confirmed it thereby. And indeed he made such advances that he soon outstripped his tutor, tho' so considerable a man." Newton, Stukeley continued, possessed an intuitive grasp of mathematics; he "wanted only a little time to maturate and deliver them. His tutor [Barrow] saw all this very plainly, conceived the highest opinion and early prognostic of his excellence; [and] would frequently say that truly he himself knew something of the mathematics, still he reckoned himself but a child in comparison of his pupil Newton."[5]

Newton's mentor and friend Isaac Barrow (1630–1677), Lucasian Professor of Mathematics at Cambridge University, proved crucial to his mathematical creativity as well as his career. Portrait by David Loggan, 1676. – National Portrait Gallery, London

Newton was undoubtedly better equipped than most to make the transition to the scientific domain. Under Henry Stokes at

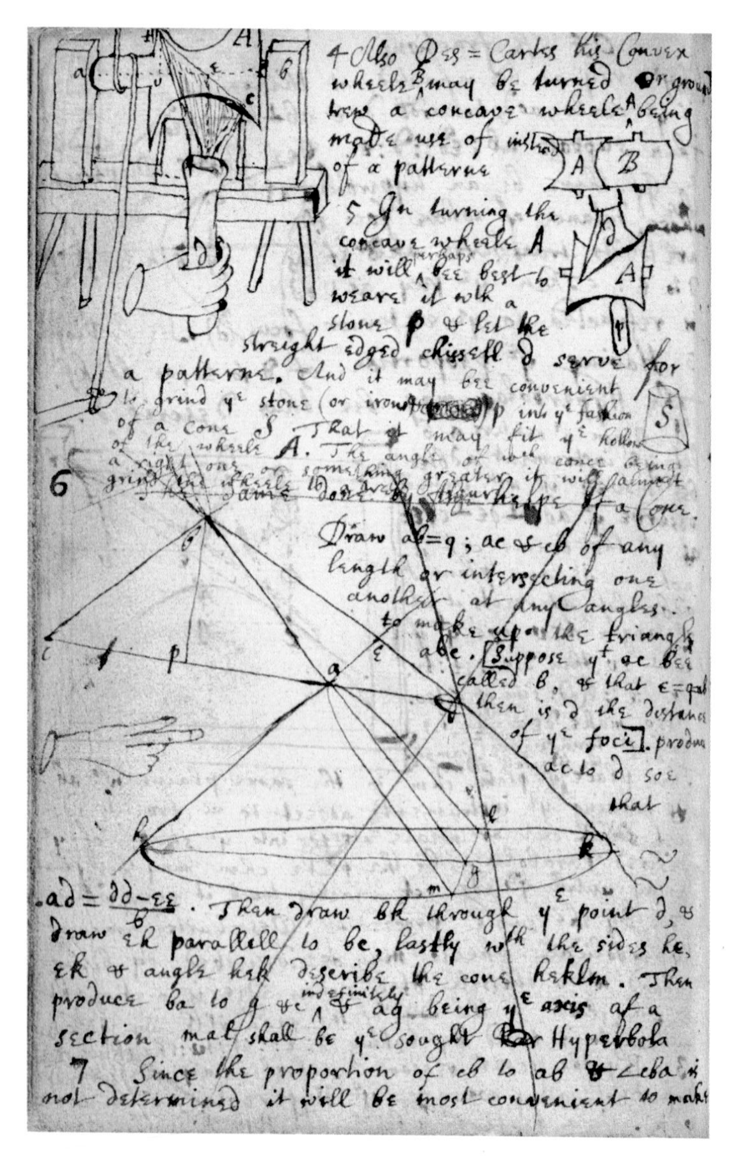

Newton's artistic and mechanical ingenuity manifested itself early, presaging the experimental scientist to come. This drawing of an apparatus for the grinding of hyperbolic lenses is taken from one of his undergraduate notebooks. – Cambridge University Library, MS ADD 4000, fol. 26v

Grantham, he had not only received the prerequisite philological and literary grounding, but had been initiated into elementary mathematics. More important were Newton's innate mechanical talents, which manifested themselves early. In school he was reputed to have "busied himself in making knick-knacks and models of wood in many kinds." These included a water clock, a couple of dials, and a wooden replica of a watermill erected near Grantham. Newton allegedly tied a mouse by the tail to the wheel and either through coercion or temptation caused the rodent to charge ahead and turn the wheel. For his schoolmates, Newton constructed kites and other amusements; for the girlfriends of young Catherine Storey – the reputed object of his early affection – he made "little tables, cupboards, and other utensils."[6]

The skills that amazed neighbors, and gave Newton a reputation of sorts in school, proved grist for the mature experimental philosopher in the making. Stukeley elaborated on the pregnant significance of Newton's artistic and mechanical talents. His "early use and expertness at his mechanical tools," the biographer wrote, "and his faculty of drawing and designing, were of service for him, in his experimental way of philosophy; and prepared for him a solid foundation to exercise his strong reasoning facultys upon."[7] Word of his intellectual promise and mechanical ingenuity undoubtedly preceded Newton to Cambridge in the form of the written (and personal) testimonials of Stokes and Babington. Nor did it take long for Newton to make good on these early prognostications.

First, however, there was the matter of the common course of studies; indeed, for the first two and a half years, Newton applied himself – much in the manner of all Cambridge undergraduates – almost exclusively to the arts of discourse and scholastic philosophy. Grounding in logic came by way of college lectures and tutorials as well as by wading through Aristotle's *Organon* ("On Logic") – partly in the original, to simultaneously improve his Greek – and Robert Sanderson's *Logicae artis compendium* ("An Epitome of the Art of Logic"). The latter text served Newton well a quarter of a century later when he came to formulate the *Principia*'s "Rules of Reasoning in Philosophy." Scholastic natural philosophy was imbibed, again by tutorials and lectures and by reading, this time

Johannes Magirus's venerable *Physiologiae peripateticae libri sex* ("Peripatetic Natural Philosophy in Six Books") and Daniel Stahl's *Regulae philosophicae* ("Philosophical Rules"). Tellingly, while Newton retained his copy of Sanderson, he felt no compunction to do the same with either Magirus or Stahl, whose usefulness evidently ended the moment he was done with them. Nor did he feel the need to retain his copy of *Ethica sive summa moralis disciplinae* ("Ethics, or a Complete System of Moral Philosophy") by Eustachius of St. Paul, his scholastic guide to moral philosophy. On that subject, Newton also read, again partly in Greek, Aristotle's *Nicomachean Ethics*. By late 1663 or early 1664, however, the new science makes its first appearance in one of Newton's undergraduate notebooks.[8]

This leap is less radical than it appears. By the second half of the seventeenth century, educators, including even those unconvinced that Cartesianism, Baconianism, and other new philosophies were superior to scholasticism, expected that the undergraduate would shift gears after he had acquired grounding in Aristotelian philosophy – deemed by all to be the prerequisite for the "modern" worldviews. The rationale, as articulated by a Cambridge champion of the new learning in the 1630s, was clear: Aristotle "hath made all learning beholding to him: no man hath learned to confute him, but by him, and unlesse hee hath plowed with his heyfer." Newton, too, at this stage in his career, regarded the new philosophy as a natural progression from his former course of study. He skipped a few pages in his notebook – perhaps anticipating a need to return at a later date to earlier topics – and simply jotted down a new heading: *Quaestiones quaedam philosophiae* ("certain philosophical questions"). Availing himself of the traditional commonplace method of note taking, he compiled a list of some forty-five topics he expected to treat. As it happened,

This depiction of the young Newton building his windmill was included in an early nineteenth-century account of Newtonian natural philosophy aimed at young readers. From Tom Telescope [John Newbery], *The Newtonian Philosophy, and Natural Philosophy in General: Explained and Illustrated by Familiar Objects in a Series of Entertaining Lectures by Tom Telescope* (London, 1838). – Burndy Library

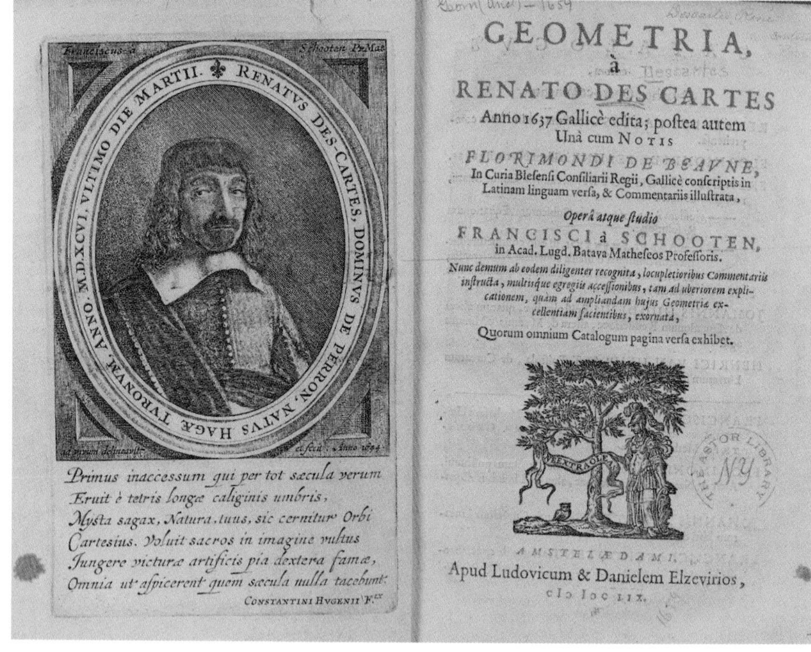

RENATVS DES-CARTES, DOMINVS DE PERRON, NATVS HAGÆ TVRONVM ANNO M.DXCVI. VLTIMO DIE MARTII.

Primus inaccessum qui per tot sæcula verum
Eruit è tetris longæ caliginis umbris,
Mysta sagax, Natura, tuus, sic cernitur Orbi
Cartesius. Voluit sacros in imagine vultus
Jungere victuræ artificis pia dextera famæ,
Omnia ut aspicerent quem sæcula nulla tacebunt.
CONSTANTINI HVGENII F.

GEOMETRIA,
à
RENATO DES CARTES
Anno 1637 Gallicè edita; postea autem
Unà cum NOTIS
FLORIMONDI DE BEAVNE,
In Curia Blesensi Consiliarii Regii, Gallicè conscriptis in Latinam linguam versa, & Commentariis illustrata,
Operâ atque studio
FRANCISCI à SCHOOTEN,
in Acad. Lugd. Batava Matheseos Professoris.
Nunc demum ab eodem diligenter recognita, locupletioribus Commentariis instructa, multisque egregiis accessionibus, tam ad uberiorem explicationem, quàm ad ampliandam hujus Geometriæ excellentiam facientibus, exornata,
Quorum omnium Catalogum pagina versa exhibet.

AMSTELÆDAMI,
Apud Ludovicum & Danielem Elzevirios,
CIↃ IↃC LIX.

Newton's thorough grasp of René Descartes' *Geometria* (Amsterdam, 1659–61) marked his progress in higher mathematics and toward the invention of the calculus.
– NYPL-SIBL

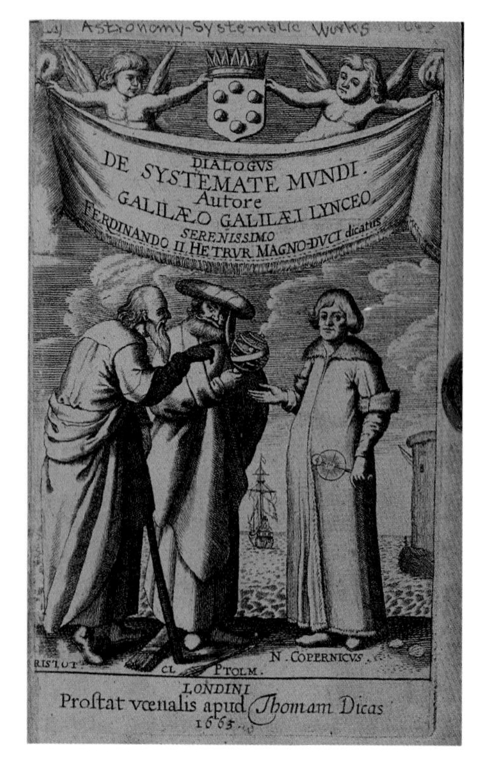

Galileo's 1632 treatise, *Dialogo sopra i due massimi sistemi del mondo*, was one of the imposing contemporary works that Newton set out to transform. Newton read the Latin translation, *Systema cosmicum*, probably in the 1663 London edition shown here.
– NYPL-SIBL

some topics received no entries, while others spawned minor essays. Later he revisited the heading and, in an act that further substantiates his perception of his new preoccupations as a natural extension of his earlier musings, he embellished it with the slogan: "Amicus Plato, Amicus Aristoteles magis amica veritas" ("Plato is a friend, Aristotle is a friend, but truth is a greater friend").[9]

Newton encountered this slogan in the very book that set him off on the new course, Walter Charleton's *Physiologia Epicuro-Gassendo-Charletoniana: Or a Fabrick of Science Natural, upon the Hypothesis of Atoms*, published in 1654. The book not only supplied Newton with a comprehensive account of ancient atomic theories, in addition to the atomic theory of the Frenchman Pierre Gassendi; it also served as a gateway to the contemporary treatises that Newton soon tracked down himself: Sir Kenelm Digby's *Two Treatises*; Galileo's *Dialogo sopra i due massimi sistemi del mondo tolemaico, e copernicano* ("Dialogue Concerning the Two Chief World Systems, Ptolemaic & Copernican"); and, most important of all, the works of Descartes. To these, Newton added Thomas Hobbes's *De corpore* ("On Bodies") and several works by Robert Boyle. In keeping with the prevailing pedagogical approach of grafting the most recent contributions onto the traditional framework of natural philosophy, the subject matter explored by Newton ranged from cosmology and matter theory to optics and meteorology, from vegetation and minerals to the soul. Understandably, Newton's penetrating intellect soon became impatient with mere note taking, substituting instead probing reflections on the issues he encountered in his readings. True, one finds little that is revolutionary in this early notebook; what does emerge is a direct correlation between topics Newton deemed worthy of note and topics he set out to transform during the seminal two years that followed.

Hand in hand with this shift in his reading, Newton also launched his career as an experimenter, in optics in particular. His scrutiny of Charleton, Descartes, and Boyle furnished him with competing theories of the new mechanical philosophy to test and challenge. His systematic prismatic experiments were yet to come, but he found his body, his eyes in particular, ideal for experimentation. His early papers demonstrate to what extremes his enthusiasm led him. He began by pressing his fingers against his eyeballs to induce colors and other effects. Inspired by the results, he proceeded to place a bodkin between his eye and the bone, "as neare to the backside of my eye as I could," and to press so as to induce curvature in the eye; "severall white darke and coloured circles" appeared when he rubbed his eye, he discovered, but not when he held the bodkin still and pressed. On another occasion, when Newton sought to explore the relationship between vision, sensations, and will, he stared at the sun for a short while through a telescope, then blinked in a dark room in order "to observe the impression made and the circles of colours which encompassed it and how they decayed by degrees." Not satisfied to carry out the experiment once, he repeated it twice more, discovering in so doing that he could recall the "phantasm" at will without looking at the sun – of course, after having locked himself in a dark room for three days to regain sight in his right eye![10]

The appearance of a comet in late 1664 served to further expand Newton's interests. On his first outing, the inexperienced Newton confused the comet with another object; a week later he got it straight. In the month that followed, he observed the comet ten more times, while simultaneously broadening his theoretical knowledge of comets by reading Willebrord Snel's *Descriptio cometae* ("Description of a Comet") and Vincent Wing's *Harmonicon celeste* ("Celestial Harmony"). With the disappearance of the comet, his interest waned; the phenomenon, nevertheless, prompted him to follow astronomy more closely. To the previous books about comets, he added Thomas Streete's *Astronomia Carolina: A New Theorie of the Coelestial Motions*, which gave him entry into the methodology of determining positions of celestial bodies as well as an acquaintance with the orbits of Johannes Kepler.[11]

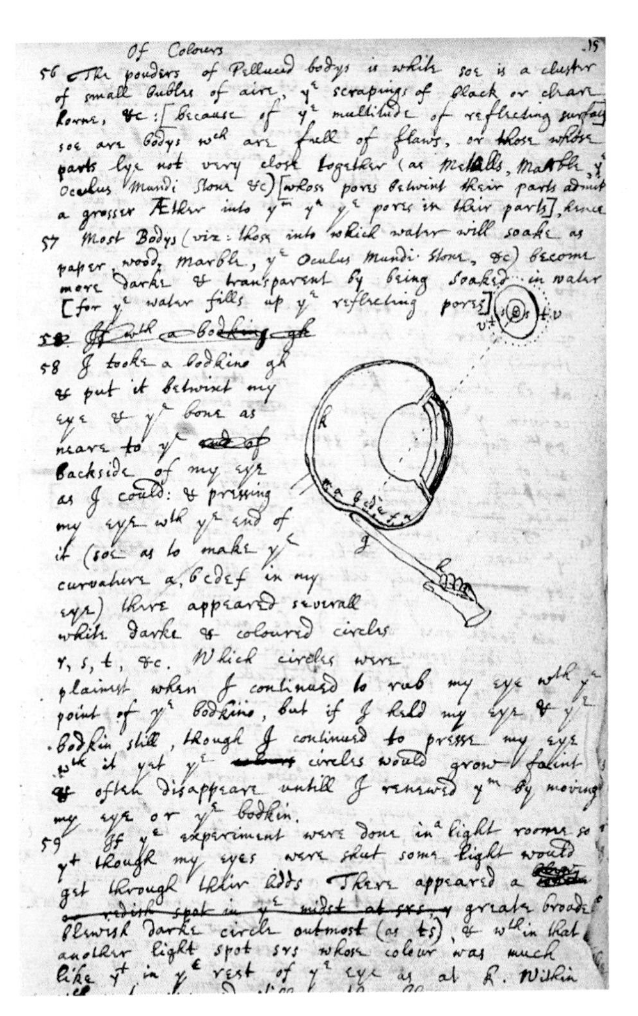

Of Colours 15

56 The pouders of Pellucid bodys is white soe is a cluster of small bubles of aire, ye scrapings of black or charre horne, &c: [because of ye multitude of reflecting surfaces] soe are bodys wch are full of flaws, or those whose parts lye not very close together (as metalls, marble, ye Oculus Mundi stone &c) [whose pores betwixt their parts admit a grosser Æther into ym yn ye pores in their parts], hence

57 Most Bodys (viz: those into which water will soake as paper, wood, marble, ye Oculus Mundi stone, &c) become more darke & transparent by being soaked in water [for ye water fills up ye reflecting pores]

58 I tooke a bodkine gh & put it betwixt my eye & ye bone as neare to ye backside of my eye as I could: & pressing my eye wth ye end of it (soe as to make ye curvature a, bcdef in my eye) there appeared severall white darke & coloured circles r, s, t, &c. Which circles were plainest when I continued to rub my eye wth ye point of ye bodkine, but if I held my eye & ye bodkin still, though I continued to presse my eye wth it yet ye circles would grow faint & often disappeare untill I renewed ym by moving my eye or ye bodkin.

59 If ye experiment were done in a light roome so yt though my eyes were shut some light would get through their lidds There appeared a greate broade blewish darke circle outmost (as ts), & wthin that another light spot srs whose colour was much like yt in ye rest of ye eye as at k. Within

In one of his early optical experiments, Newton inserted a bodkin between his eye and his skull bone. He described and illustrated the experiment in one of his notebooks (Cambridge, 1660s). – Cambridge University Library, MS ADD 3975, p. 15

Chances are that Newton's independent course of reading – if not experimentation – was undertaken with the encouragement of a kindred spirit. Certainly, Newton owned none of the books he read during this period, which leads one to speculate that the well-furnished library of Isaac Barrow was the source of his philosophical texts as it had been earlier for his mathematical texts. Quite likely, too, the older scholar, who had himself followed an almost identical educational path a decade earlier, was in a position to resolve whatever problems or doubts Newton may have encountered. Certainly, by the time Newton made Charleton, Gassendi, and Descartes his companions, Barrow was playing a key role in advancing Newton's mathematical training as well as in helping secure the young undergraduate a Trinity College scholarship.

In old age, Newton told his admirers that he had commenced his mathematical studies by chance. While visiting Sturbridge Fair in summer 1663, he had purchased an astrological treatise wherein he met with a "figure of the heavens" that confounded him, owing to his want of trigonometry. He promptly acquired a suitable textbook, but found its demonstrations, too, beyond his ken. So he returned to the basics and purchased a copy of Euclid's *Elements* – Barrow's 1655 edition. A cursory reading of "the titles of the propositions, which he found so easy to understand at first," made him dismissive of the book: "he wondered how any body would amuse themselves" to write such easy demonstrations. He changed his mind when he found himself struggling with the reasoning behind the Pythagorean theorem and "began again to read Euclid with more attention than he had done before and went through it." Sometime later, he claimed, he supplemented Euclid with Oughtred's *Clavis mathematicae* ("The Key to Mathematics") and Descartes' *La Géometrie* ("Geometry"). According to another anecdote, when Newton stood candidate for the Trinity scholarship in late April 1664, his mastery of Euclid was still somewhat shaky, and this failing nearly cost him the election. Barrow was the examiner, and he proceeded to scrutinize Newton's grasp of Euclid, which, Newton recalled more than half a century later, he had neglected and "knew little or nothing of." For his part, Barrow "never asked him about Descartes's Geometry which he [Newton] was master of." Naturally, so the story goes, the candidate was too modest to brag about his proficiency and since Barrow "could not imagine that anyone could have read that book without first being master of Euclid," he formed a rather "indifferent opinion" of Newton, who, nonetheless, got his scholarship.[12]

Both stories are brazen embellishments. Newton's knowledge of Euclid in early 1664 was perhaps more profound than he led his admirers to believe, while his mastery of Descartes was still a thing of the future. Indeed, he himself recorded in 1699 the precise chronology of his mathematical studies: "By consulting an acompt of my expenses at Cambridge ... I find that in the year 1664 a little before Christmas I being then senior Sophister, I bought Schooten's Miscellanies and Cartes's Geometry (having read this geometry and Oughtred's Clavis above half a year before)."[13] His initial encounter with Descartes and Oughtred, then, followed his election to a scholarship some eight months earlier; he could scarcely have been "master" when examined by Barrow. But it was a good story with which to regale his admirers.

It was almost certainly the investiture of the Lucasian lectures that kindled Newton's passion for higher mathematics. Barrow delivered his inaugural oration on March 14, 1664, two weeks before the Trinity election, and Newton was undoubtedly present. Years later, he acknowledged having attended the lectures, further admitting that these may have helped inspire his inception of the calculus: "Dr Barrow then read his Lectures about motion and that might [have] put me upon taking these things into consideration."[14] It has been suggested that the third lecture, in particular, wherein Barrow discussed the limit-sum of a converging geometrical progression, could possibly have been the source from which Newton later drew his fundamental statement of differentiation. Newton did not elaborate, and in any case he was sparing in his references to his former mentor: by the turn of the eighteenth century Barrow had become a key player in the celebrated priority dispute with Leibniz over the invention of the calculus, with each camp accusing the other of appropriating from Barrow.[15] Be that as it may, once Newton embarked on higher mathematics, he made rapid progress. He digested – fully and thoroughly – Frans van Schooten's second Latin edition of Descartes' *Géometrie*; he perused François Viète's *Opera mathematica* ("Mathematical Works"; also edited by Schooten); he carefully studied John Wallis's *Arithmetica infinitorum* ("Arithmetic of Infinites") and the *Commercium epistolicum*. Then the plague struck Cambridge.

A common perception lingers that with the outbreak of the epidemic in 1665, Newton left Cambridge for two years and that it was precisely during these *anni mirabiles* – two wonder years of isolation – that Newton made his great discoveries. In actual fact, Newton was away for only two periods of eight or nine months each, with a creative Cambridge interval in between. Having left for Lincolnshire in June 1665, Newton was back in residence by mid-March 1666, remaining in Cambridge until mid-June. He then left again, not to return until April 1667. Newton's subsequent chronology of his discoveries allows us to c orrelate the progress of his studies with his whereabouts.

As Newton recollected, he discovered "the Method of approximating series and the Rule for reducing any dignity of any Binomial into such a series" in early 1665, and in May he

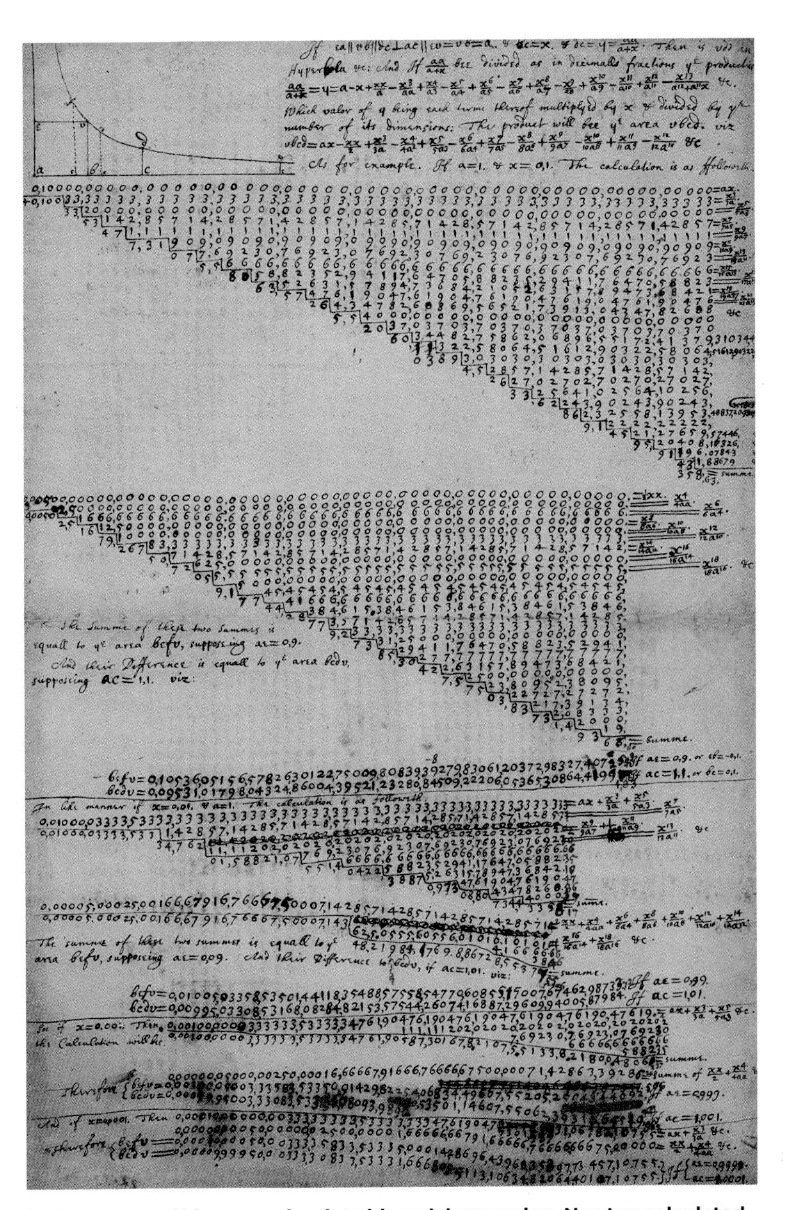

In the course of his researches into binomial expansion, Newton calculated (ca. 1665) several logarithms up to fifty-five places. – Cambridge University Library, MS ADD 3958, fol. 79r

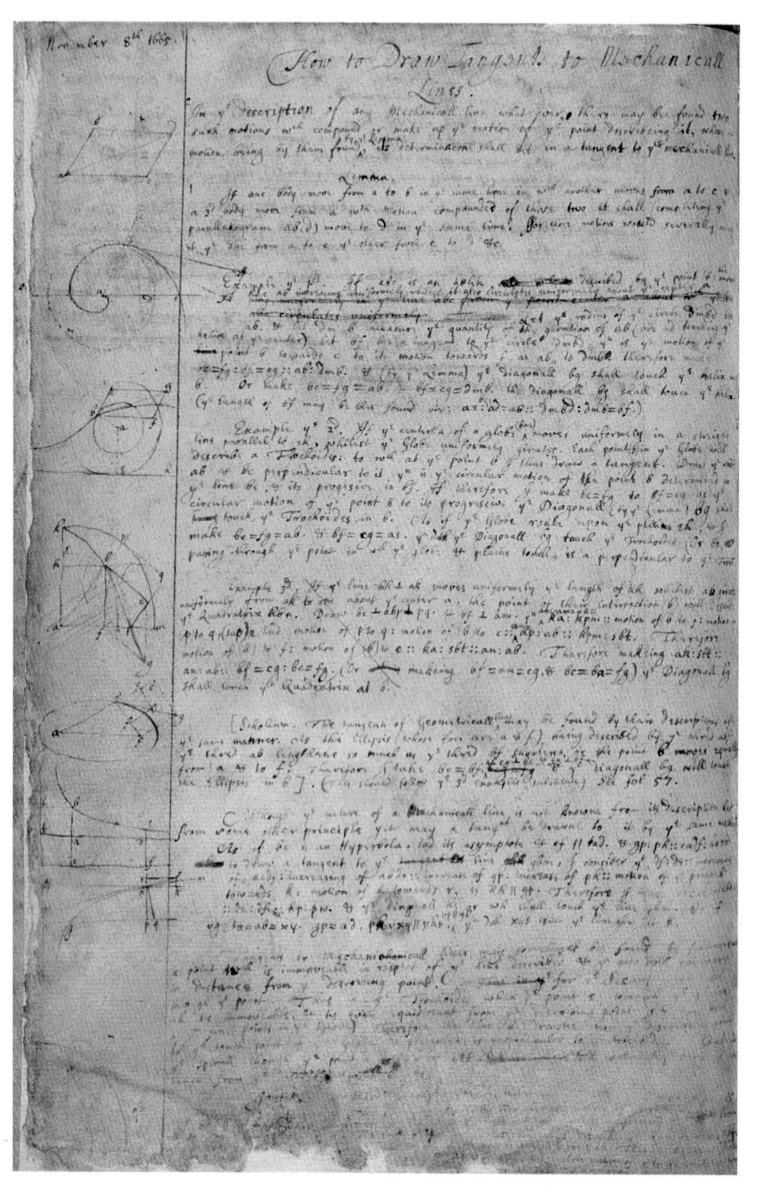
November 8th 1665.

How to Draw Tangents to Mechanicall Lines.

The first page of Newton's "How to Draw Tangents to Mechanicall Lines" (November 8, 1665), from his "Waste Book." – Cambridge University Library, MS ADD 4004, fol. 50v

came by "the method of Tangents." He left for the country in June 1665, shortly after Barrow had finished delivering the first set of his geometrical lectures, supplemented by lectures on Archimedes. Newton then devoted several months to synthesizing his earlier discoveries; by November, he "had the direct method of fluxions" – the Newtonian form of the infinitesimal calculus. Then, "the light went out." Newton set mathematics aside and turned to optics, claiming to have discovered the theory of colors in January 1666. His return to Cambridge appears to have revitalized his mathematical interests, for during his three-month stay at Trinity – where he also found Barrow, about to deliver his last set of mathematical lectures – Newton "had entrance into the inverse method of fluxions." Returning to Lincolnshire, Newton digested the new material and, in an inspired burst of creativity in October, composed his foundational paper on the resolution of problems by motion. Newton then cast mathematics aside for two years. In this light, then, the Cambridge environment, specifically his access to books as well as to Isaac Barrow, was crucial to his mathematical creativity. As the editor of Newton's mathematical papers concluded, most of them were written while he was in residence, not in Lincolnshire, but in Cambridge, while "those written in the country were revised, polished accounts of researches originally pursued at university."[16]

While Newton's claims for full-grown maturity in the domain of mathematics during the plague years are borne out by his papers, his achievements in other domains fell shortof what he subsequently insinuated. In addition to the theory of colors, Newton pinpointed 1666 as the year in which he

> began to think of gravity extending to the orb of the Moon and (having found out how to estimate the force with which [a] globe revolving within a sphere presses the

surface of the sphere) from Keplers rule of the periodical times of the Planets being in a sesquialterate proportion of their distances from the center of their Orbs, I deduced that the forces which keep the Planets in their Orbs must [be] reciprocally as the squares of their distances from the centers about which they revolve: and thereby compared the force requisite to keep the Moon in her Orb with the force of gravity at the surface of the earth, and found them answer pretty nearly. All this was in the two plague years of 1665–1666. For in those days I was in the prime of my age for invention and minded Mathematicks and Philosophy more then at any time since.[17]

In his old age, Newton claimed that he discovered universal gravitation in a flash upon seeing an apple falling in his mother's garden at Woolsthorpe. In fact, it took him two decades to work out the insights of 1666. – Burndy Library

The implication that universal gravitation struck him early and, as it were, in a "flash" was embellished by Newton shortly before his death when he was asked about the origins of his great discovery. Both Stukeley and John Conduitt – the husband of Newton's niece Catherine Barton – were told that it was the sight of a falling apple in his mother's garden that set Newton musing "that the power of gravity ... was not limited to a certain distance from the earth but that this power must extend much farther than was usually thought. Why not as high as the moon ... and if so that must influence her motion and perhaps retain her in her orbit." Proceeding to calculate, he found a fairly close correlation between the force that kept the moon in its orbit and the force of gravity. We now know that Newton, in the words of Richard S. Westfall, did not carry the *Principia* "about with him essentially complete for twenty years until Halley pried it loose and gave it to the world." Certainly, he made impressive progress in applying the new mathematics to the investigation of orbital trajectories. Yet in his youth he never availed himself of Kepler's area law (stating that planetary orbits sweep out equal areas in equal times); if anything, he had only begun to explore the nature of "mass" and "force." Likewise, Newton lacked in 1666 much of the empirical basis upon which the propositions of the *Principia* were to be established. Working out these issues preoccupied him for two decades, and required, in addition to his legendary "patient thought" and mathematical prowess, the crucial contributions of Christiaan Huygens, Robert Hooke, and John Flamsteed; it would also be necessary for him to extricate himself from the last vestiges of Cartesian influence, a process that was completed only when he wrestled with the *Principia*.[18]

Be this as it may, within two years of returning to Cambridge, Newton's career changed drastically. For some time, Barrow had been contemplating relinquishing his professorship; for all his mathematical and philological prowess, Barrow never wavered in the belief that his true calling was as a divine. Recognizing now that he had a more than worthy successor waiting in the wings, Barrow resolved to hand over the professorship, which all along he had seen himself filling as a caretaker of sorts. Before 1669, however, his hands were tied. The statutes required not only that the Lucasian Professor

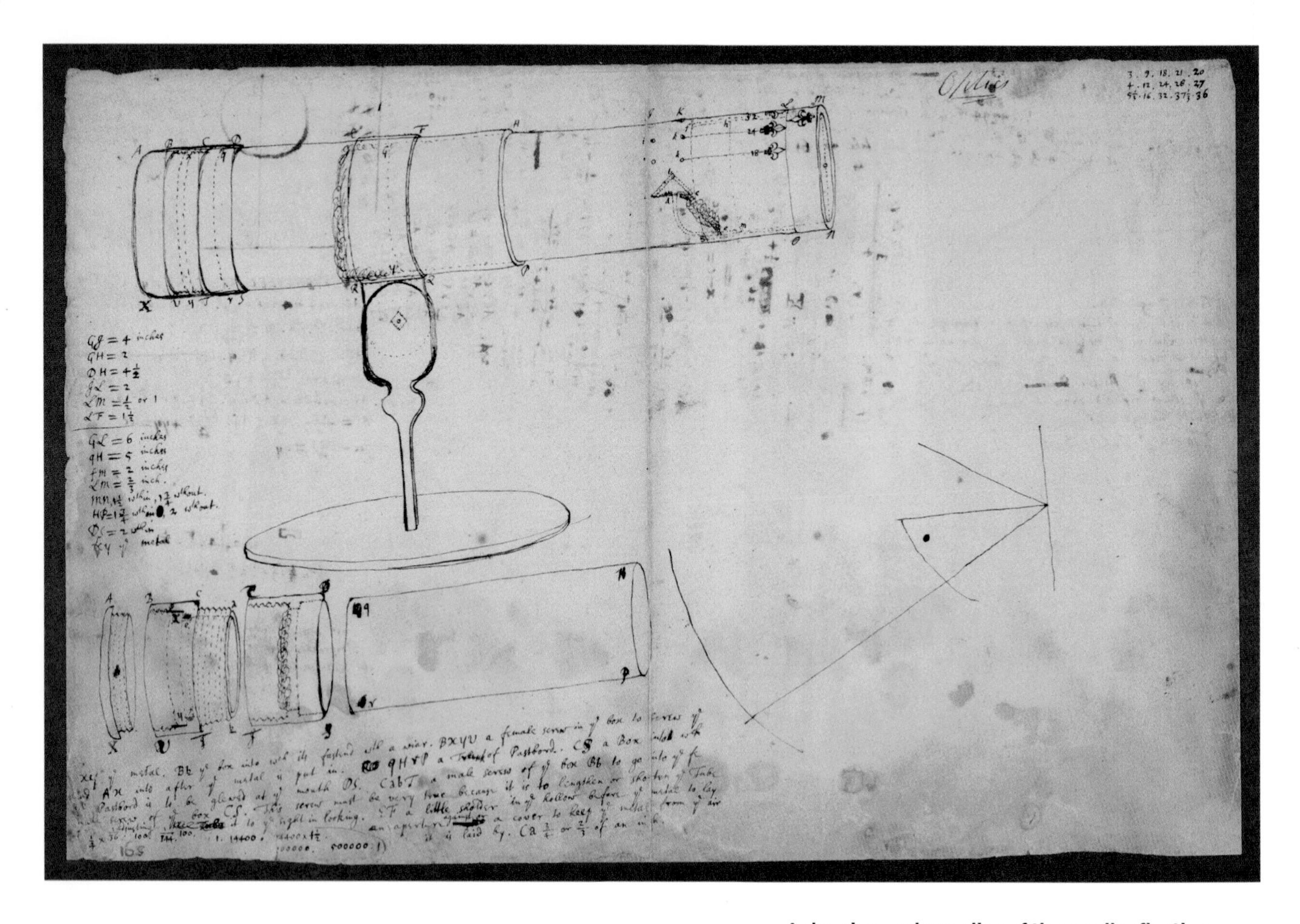

A drawing and a replica of the small reflecting telescope that catapulted Newton to European fame. Invented in 1668, the telescope measured a mere seven inches in length, with an aperture slightly larger than an inch. – Drawing: Cambridge University Library, MS ADD 3970, fols. 591r–592v; replica (wood and paper): Courtesy of Adler Planetarium & Astronomy Museum, Chicago, Illinois

be a Master of Arts at the very least, but that he be of proven erudition as well. Newton graduated M.A. on July 7, 1668, and turned twenty-six – the minimum age stipulated in the Savilian statutes, upon which the Lucasian statutes were modeled – the following Christmas Day. All that was required now was a public demonstration of Newton's talents. The opportunity presented itself in early 1669 when Barrow encouraged his protégé to compose his *De analysi per aequationes numeri terminorum infinitas* ("On Analysis by Infinite Series"). This treatise was expected both to assert Newton's priority in discovering a general method of infinite series (and of the method of fluxions more generally) – necessitated by the recent publication of Nicholas Mercator's *Logarithmotechnia* ("The Art of Constructing Logarithms") – and to demonstrate to a wider audience his impeccable credentials. Newton complied. In July 1669, the paper was sent, first to the mathematician John Collins and then to William, Lord Brouncker, president of the Royal Society, and to John Wallis, Savilian Professor of Geometry. With the accolades and endorsement of the Royal Society mathematicians in hand, on October 29, 1669, Barrow easily convinced the trustees of the Lucasian professorship to appoint Newton as his successor.

Newton paid the proper tribute owed benefactors on such occasions: he opted to begin his tenure by lecturing on optics, the last subject treated by Barrow. For his part, Barrow remained steadfast in his determination to make Newton's genius public. In December 1671, Barrow, then royal chaplain, persuaded Newton to allow him to present King Charles II with the reflecting telescope that Newton had constructed. It was this small instrument that catapulted the newly elected professor to European fame. The origins of the telescope can be pinpointed to 1668, when Newton's renewed researches into refraction led him to construct a model measuring a mere six inches in length, with an aperture slightly larger than an inch, "and a Plano-convex eye glasse whose depth is 1/6th or 1/7th part of an Inch." Such a tiny instrument, its inventor boasted, not only eliminated the chromatic aberration that bedeviled telescopes with glass lenses, but it magnified "about 40 times in Diameter which is more than any 6 foote Tube can doe."[19] For some reason, Newton set his early model aside; only in 1671 did he build another, slightly larger telescope, measuring some seven inches in length and two and a quarter inches in diameter; it was this instrument that Barrow exhibited to the Royal Society. Newton's exposure to the London philosophers who comprised the Society, and especially to its Curator of Experiments, the physicist Robert Hooke, was destined to figure significantly in his scientific career.

The optical part of Hooke's *Micrographia* played for Newton a role analogous to that of Descartes' *Principia* – that is, it served as an abundant and challenging resource that Newton set out to destroy, or at least radically transform. Newton probably borrowed a copy of the book before leaving Cambridge in June 1665, and his surviving notes make it plain that he reacted strongly to Hooke's views. To Hooke's claim that "Light is a vibration of the Aether, which pulse is made oblique by refraction," for example, Newton retorted: "Why then may not light deflect from streight lines as well as sounds?" He liked even less Hooke's explanation that there existed only two primary colors, which are modifications of white light: blue, which "is an impression on the Retina of an oblique and confused pulse of light, whose weakest part precedes, and whose strongest follows"; and red, an impression "whose strongest part precedes." But as flawed as his theory may have been, it was Hooke who provided the most comprehensive mechanist account of the various forms generating colors, and who raised the bar by stipulating that a true theory of colors

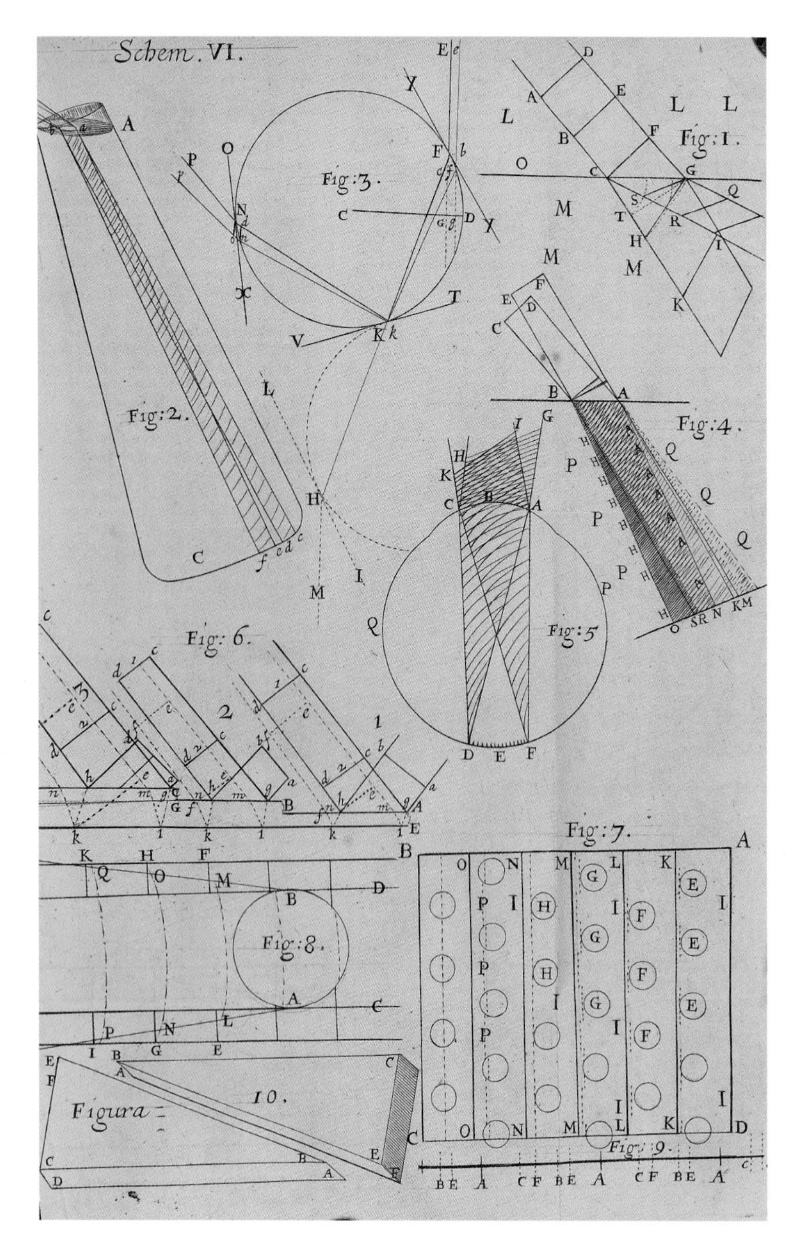

Robert Hooke's research on optics served as an abundant and challenging resource that Newton set out to radically transform. This is a plate from Hooke's *Micrographia: or, Some Physiological Descriptions of Minute Bodies Made by Magnifying Glasses* (London, 1665). – NYPL–Rare Books Division

must encompass all of them. Newton – for all his ire – was profoundly influenced by the *Micrographia*, "more than he was ever able to admit."[20]

Newton was thrust into the international scientific community during the closing days of 1671, following Barrow's delivery of the remarkable reflecting telescope into the hands of the Royal Society. The members expressed their appreciation by proceeding to elect the Cambridge professor a Fellow. So overwhelmed was Newton by the reception of his "toy" that he immediately addressed a clumsy letter to Henry Oldenburg, Secretary of the Royal Society, in which he proposed to send the Society, "to be considered of and examined" by it, an account of "a Philosophicall discovery" that, in his estimation, was "the oddest if not the most considerable detection which hath hitherto beene made in the operation of Nature." We can only speculate on the expectations raised by such a promise when the letter was read at the meeting of January 25, 1672. Newton did not keep them long in suspense. He dispatched his account of light and colors on February 6, and two days later it reached Oldenburg. Newton was later informed that it was read that very afternoon, and greeted "with a singular attention and an uncommon applause." The members also expressed the desire that it be printed immediately, lest "the ingenuous and surprising notion therein contained" be "snatched" away from Newton.[21]

The Society appointed three Fellows to scrutinize Newton's theory of the unequal refrangibility of light: Seth Ward, Robert Boyle, and Robert Hooke. As often happened in such cases, Hooke alone discharged his duty, producing his report at the Society's very next meeting. Meanwhile, Oldenburg, having received Newton's consent for publication, rushed the paper to press – even before the Society reconvened. In fact, except for two book reviews, issue number 80 of the Society's *Philosophical Transactions* was entirely devoted to the "New Theory About Light and Colors." Hence, it must have been something of a shock and embarrassment to Secretary Oldenburg to hear Hooke propounding his critique of Newton's theory on February 15.[22]

Much abuse has been showered on Hooke for this critique: a senior member of the scientific community, Hooke peremptorily dismissed an original paper by a young researcher for no reason other than to defend his own theory. Closer examination of this "scandalous" chapter in the history of science, however, somewhat redeems the vilified Hooke. On the simplest level, the twenty-nine-year-old Cambridge Professor of Mathematics was not all that junior – either in status or in accomplishment – to the thirty-six-year-old Curator of the Royal Society. More important, while Hooke was certainly fond of his own hypothesis, his critique cannot be attributed solely to partisanship. His job as designated referee was to subject *any* report to scrutiny, and thereby facilitate debate within the Society. That Hooke focused on a limited range of topics – and in retrospect, not necessarily the important ones – was a function of what he considered particularly unsettling in Newton's paper: not the verity of Newton's prismatic experiments, which he accepted, but certain problematic methodological and conceptual issues – for example, Newton's refusal to offer a physical explanation for the phenomena he described. It was precisely this refusal that troubled *all* early critics of the theory, and the cumulative effect of their uneasiness prompted Newton in 1675 to produce his "hypothesis explaining the Properties of Light."

Also troubling, not just for Hooke but for other contemporaries, was the inherent dogmatism of Newton's claim that his theory was established according to the "most rigid consequences" (the Dutch mathematician and physicist Christiaan Huygens believed, as he wrote Oldenburg, that the "thing could very well be otherwise," and that Newton "ought to content himself if what he has advanced is accepted as a very likely hypothesis"). More egregious still, as far as Hooke was concerned, was Newton's covert disparagement of the naturalist and experimental traditions of the Royal Society, evident in the younger man's trumpeting that "a naturalist would scearce expect to see the science of those become mathematicall, and yet I dare affirm that there is as much certainty in it as in any other part of Opticks." Hooke was probably the only person (other than Oldenburg) to read this paragraph; the Secretary was careful to excise it from the published paper.[23]

This background makes it easier to comprehend Hooke's reaction. On the one hand, he endorsed Newton's experiments as valid, "having by many hundreds of tryalls found them soe." On the other hand, as the mouthpiece of an institution that explicitly shunned hypotheses, Hooke admonished Newton's presumption to derive mathematical demonstrativeness from experiments. More than simply defending his own theory, Hooke imparted a methodological message: the experiments warranted neither Newton's "hypothesis of Light" nor even his own, for the phenomena could be explained equally well according to two or three *other* theories.

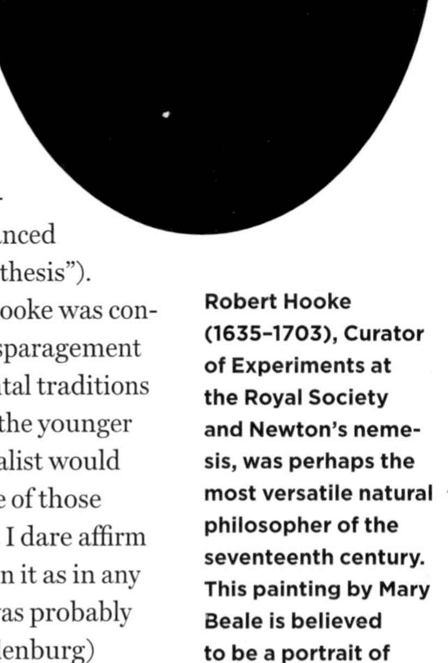

Robert Hooke (1635–1703), Curator of Experiments at the Royal Society and Newton's nemesis, was perhaps the most versatile natural philosopher of the seventeenth century. This painting by Mary Beale is believed to be a portrait of Hooke. – © The Natural History Museum, London

In the months that followed, Hooke and Newton continued to harp on the relations between theory and experiment, while Oldenburg proceeded to expurgate Newton's responses before their publication in the *Philosophical Transactions*. Hooke's rebuttals did not matter, for Oldenburg never bothered to print them.[24]

Oldenburg's early and unwavering partisanship of Newton was not wholly disinterested. In the early 1670s, Oldenburg's relations with Hooke had deteriorated, and now that Newton had turned against the Curator of Experiments following the receipt of his critique, Oldenburg seemed to relish inciting discord between the two. Hooke, who replicated and defended the verity of Newton's experiments while disputing the theory, seemed unaware of Oldenburg's duplicity. Only in late 1675, when (in no small part owing to the persistent prodding of Oldenburg) Newton finally dispatched his "hypothesis of Light" – in which he savagely attacked both Hooke's optical ideas and his integrity – did the Curator surmise something was afoot. On the evening of January 20, 1676, immediately after the meeting at which Newton's letter was read to the Society, Hooke noted in his diary: Newton "seeming to quarrell from Oldenburgs fals suggestions." That same evening he also wrote Newton directly, and for the first time. Professing his utmost respect for the Lucasian Professor, he explained that his critique was made in the cause of truth, informed by the liberty allowed in philosophical matters. Newton seemed to accept the overture: he desisted from further escalating the breach with Hooke, and refused to allow publication of the "hypothesis." Henceforth, he distanced himself from Oldenburg.[25]

In fact, Newton distanced himself from the scientific community more generally. For the previous three years, he had been embroiled in debates regarding his theory of colors not only with Hooke, but with Huygens, the Parisian Jesuit Ignace-Gaston Pardies, and, in particular, a group of obdurate Jesuits at Liège. On several occasions, Newton threatened to withdraw from active participation in the republic of scientific letters – and even resign his Royal Society Fellowship. After 1676, he finally made good on his threat. His interests had shifted as well. He discovered alchemy in 1669, and the 1670s saw him plunge into major experimental activity, so much so that in November 1675, John Collins informed a friend that he had not bothered Newton for nearly a year, as the latter was "intent upon Chimicall Studies and practises, and both he and Dr Barrow ... beginning to thinke math[emati]call Speculations to grow at least nice [i.e., foolish] and dry."[26] It was Hooke who drew him back.

Henry Oldenburg (ca. 1615–1677), Secretary of the Royal Society and publisher of the *Philosophical Transactions*, was an unwavering early supporter of Newton. - © The Royal Society

On November 24, 1679, Hooke wrote Newton in order to initiate an exchange on, among other topics, planetary motion. Hooke had been hard at work on the subject for nearly two decades. He made public his important conjectures on the topic in two lectures delivered at Gresham College in 1666 and 1670 – the former of which was published in 1674 as *An Attempt to Prove the Motion of the Earth*. During the two years prior to writing Newton, he had made further progress. Now, having become Secretary of the Royal Society in 1677 following Oldenburg's death, and having more recently added the correspondence of the Society to his duties, Hooke availed himself of the opportunity to combine his own interests with those of the Society; he invited Newton to resume communication with the London philosophers. In particular, he added, "I shall take it as a great favour if you shall please to communicate by Letter your objections against any hypothesis or opinion of mine, And particularly if you will let me know your thoughts of that of compounding the celestiall motions of the planetts of a direct motion by the tangent and an attractive motion towards the centrall body."[27]

The request spawned a chain of events that culminated with the publication of the *Principia* in 1687. Newton responded immediately, though evasively, to Hooke's letter, claiming that he "shook hands with Philosophy," that he had never heard of Hooke's theory, and that he had been preoccupied with other affairs. Still, he proceeded to offer Hooke "a fansy" of his own "about discovering the earth's diurnal motion" – an experiment "concerning the descent of heavy bodies for proving the motion of the earth." The Secretary chose to ignore that Newton actually let slip his familiarity with Hooke's theory – a copy of the book was also on his shelf – in the interest of further

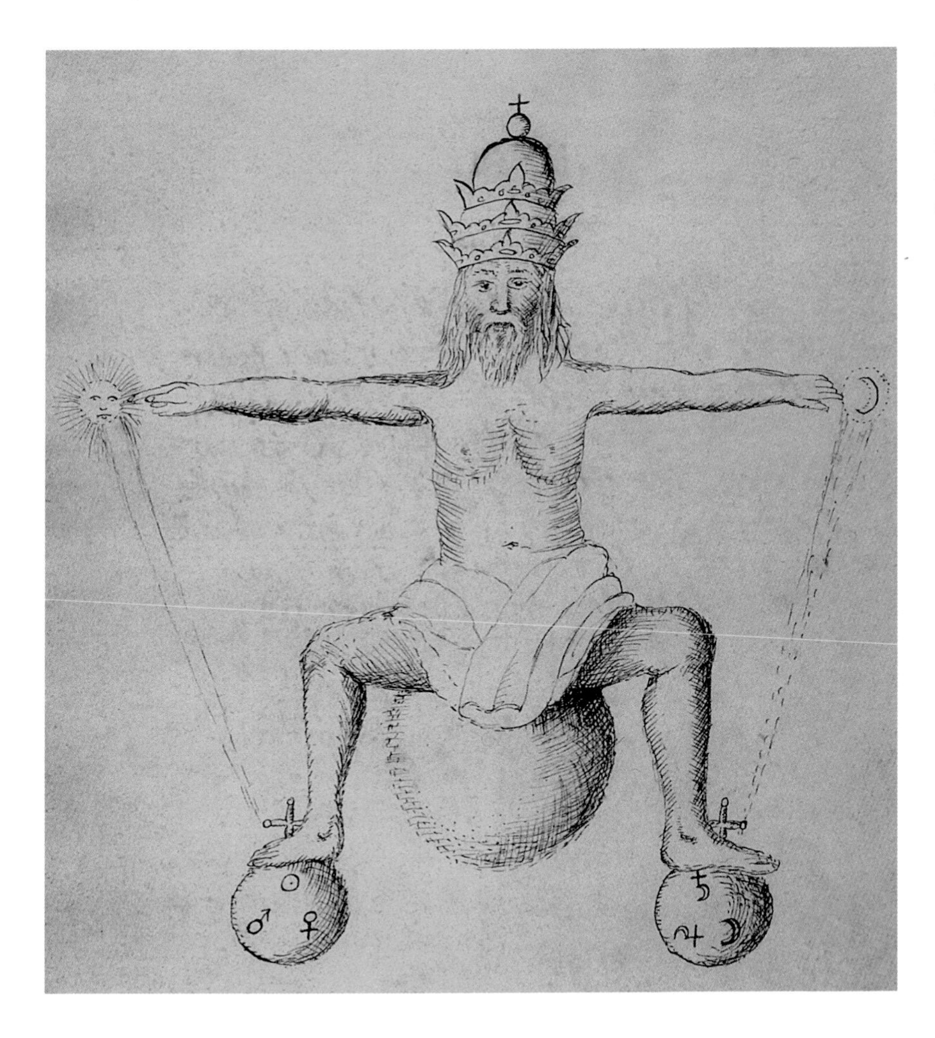

Newton's drawing of Jupiter enthroned comes from a notebook dating from the 1670s; he copied it from an anonymous English translation of Johann de Monte-Snyder's *Metamorphosis planetarum* (1663). – Yale University, Harvey Cushing/John Hay Whitney Medical Library

engaging the reluctant, but evidently not unwilling, Cambridge professor. Hooke wrote back to express agreement with Newton's proposed experiment regarding the descent of bodies as well as to point out an error. Newton's attention was piqued, as was his indignation; he resented being proven wrong, above all by Hooke. He responded, but even more guardedly than before. Hooke was not to be deterred. After a flurry of activity that lasted several weeks – and brought Hooke to "perfect [his] Theory of Heavens" – he attempted to bait Newton with more explicit details of his supposition "that the Attraction always is in a duplicate proportion to the Distance from the Center Reciprocall, and Consequently that the Velocity will be in a subduplicate proportion to the Attraction and Consequently as Kepler Supposes Reciprocall to the Distances." Newton did not respond.[28]

Ten days later, after he had performed the experiment suggested by Newton, Hooke wrote again to acquaint the Cambridge professor that such a success "will prove a Demonstration of the Diurnall motion of the earth as you have very happily intimated." All that remained, he added, was

to know the proprietys of a curve Line (not circular nor concentricall) made by a centrall attractive power which makes the velocitys of Descent from the tangent Line or equall straight motion at all Distances in a Duplicate proportion to the Distances Reciprocally taken. I doubt not but that by your excellent method you will easily find out what that Curve must be, and its proprietys, and suggest a physicall Reason of this proportion.[29]

Newton, again, remained stubbornly silent. He appears to have swiftly obtained a solution – not to Hooke's question vis-à-vis determining the shape of an orbit given a force, but to its converse – yet he never transmitted the solution to Hooke; to have done so would have been to forfeit his own earlier work, not to mention hand over to his antagonist a glorious discovery. Newton later acknowledged that Hooke's letters occasioned his "finding the method of determining Figures," though the "duplicate proportion I can affirm that I gathered it from Keplers Theorem about 20 yeares ago." Likewise, he admitted that Hooke taught him of "the deflexion of falling bodies to the south east in our Latitude" as well as brought to his attention Edmond Halley's pendulum experiments at St. Helena – though, again, Newton insisted that he had discovered the *notion* of "gravities being lessened at the equator by the diurnal motion" almost two decades earlier.[30]

In the heat of the clash over Hooke's demand that Newton acknowledge his debt to him, Newton enunciated what he regarded to be the crux of the matter, both for planetary motion and for optics: namely, that both are fundamentally *mathematical sciences*. True, Hooke's letters discussed gravity, duplicate proportion, and ellipsoidal orbits, "and this he did in such a way as if he had found out all and knew it most certainly." But Hooke's manifest ineptitude in mathematics spoke volumes against such pretensions of discovery:

he has done nothing and yet written in such a way as if he knew and had sufficiently hinted all but what remained to be determined by the drudgery of calculations and observations, excusing himself from that labour by reason of his other business: whereas he should rather have excused himself by reason of his inability. For tis plain by his words he knew not how to go about it. Now is not this very fine? Mathematicians that find out, settle and do all the business must content themselves with being nothing but dry calculators and drudges and another that does nothing but pretend and grasp at all things must carry away all the invention as well of those that were to follow him as of those that went before.[31]

Herein lies one of Newton's greatest contributions: the establishment of the physical sciences on the secured foundation of mathematics. Hence, as perceptive as Hooke's intuition might have been, and as ingenious as his experiments proved to be, without the "drudge" of calculation they did not amount to much. Newton drove the point home by drastically restructuring Book Three of the *Principia* and enshrining in its preface his views regarding the preeminence of demonstrativeness over philosophical intuition:

In the preceding books I have presented principles of philosophy that are not, however, philosophical but strictly mathematical.... It still remains for us to exhibit the system of the world from these same principles. On this subject I composed an earlier version of book 3 in popular form, so that it might be more widely read. But those who have not sufficiently grasped the principles set down here will certainly not perceive the force of the conclusions, nor will they lay aside the preconceptions to which they have become accustomed over many years; and therefore, to avoid lengthy disputations, I have translated the substance of the earlier version into propositions in a mathematical style, so that they may be read only by those who have first mastered the principles.[32]

Though the *Principia* is ostensibly devoid of epistemology, ontology, and metaphysics, it had enormous implications for Enlightenment philosophy, and not just because the critical blow it inflicted on Descartes' physics bore directly on other domains of Cartesian philosophy. The ability of Newtonian science to derive certain knowledge directly from the phenomena of nature, for example, threatened the very foundation of rationalist philosophy, based as it was on the supposition that reason alone provides the means to rise above the limitations of our senses and arrive at sound theoretical knowledge. Likewise, the incessant preoccupation of philosophy with first principles, with the causes and essences of things, was rendered irrelevant by Newton's insistence that it was sufficient to demonstrate, mathematically or experimentally, their existence. Such radical implications for the common practice of philosophy generated, as we shall see below, much of the opposition to Newtonian ideas. Newton anticipated this eventuality, but nonetheless refrained from engaging in more "traditional" philosophy – fearing, perhaps, that to do so would deflect attention from what truly mattered, and would entangle him in sterile disputes.

The *Principia* was published to universal acclaim – in England at least – in early July 1687. As we shall see below, the more protracted process of reception on the Continent – not only of the *Principia* but of the *Opticks* – was due partly to choices of style and substance required by the polemical context in which Newton seemed always to find himself. Though the *Principia* could not have been made any less demanding, Newton might have shown himself more willing to partake in the process of popularization. Instead, his position on the matter hardened. Late in life he even boasted that he had made the *Principia* deliberately difficult in order to avoid "being baited by little Smatterers in Mathematicks." Hooke, no doubt, was the smatterer in question.[33]

Back in February 1675, when Newton accepted the olive branch Hooke had extended to him, he availed himself of an already venerable metaphor: "If I have seen further it is by standing on the shoulders of Giants." Newton's striking self-image, however, is at variance with the modest spirit traditionally attributed to it. Put bluntly, Newton was not handing out

PHILOSOPHIÆ

NATURALIS

PRINCIPIA

MATHEMATICA.

Autore *J S.* NEWTON, *Trin. Coll. Cantab. Soc.* Matheſeos Profeſſore *Lucaſiano*, & Societatis Regalis Sodali.

IMPRIMATUR.

S. PEPYS, *Reg. Soc.* PRÆSES.

Julii 5. 1686.

LONDINI,

Juſſu *Societatis Regiæ* ac Typis *Joſephi Streater*. Proſtant Venales apud *Sam. Smith* ad inſignia Principis *Walliæ* in Cœmiterio D. *Pauli*, alioſq; nonnullos Bibliopolas. *Anno* MDCLXXXVII.

Newton's masterpiece, *Philosophiæ naturalis principia mathematica* (commonly known as the *Principia*), repudiated Descartes by establishing mathematical foundations for natural philosophy. Even the title page of the first edition (London, 1687) was designed to signal Newton's vanquishment of his rival: the title of Descartes' book is fully incorporated, in boldface and set in larger type, into Newton's title. – NYPL–Rare Books Division

a florid – and uncharacteristic – compliment; he was settling one last score. A month earlier, following the reading of Newton's "hypothesis of Light" at a meeting of the Royal Society, the beleaguered Hooke dismissed it indignantly: the hypothesis, he fired from his chair, "was contained in his *Micrographia*, which Mr. Newton had only carried farther in some particulars." Oldenburg, no surprise, informed the Lucasian Professor of the rejoinder. He needn't have bothered; Hooke reiterated much the same sentiment when he wrote Newton how pleased he was "to see those notions promoted and improved which I long since began, but had not time to compleat. That I judge you have gone farther in that affair much than I did." Newton's retort was as civil as it was brutal. Descartes had made a good start, he wrote, and Hooke had "added much several ways, and especially in taking the colours of thin plates into philosophical consideration." Then, taking as his cue Hooke's own words – "I judge you have gone farther" – Newton inflicted on the hunchbacked Curator that backhanded (and cruel) compliment: "If I have seen further it is by standing on the shoulders of Giants."[34]

In Newton's mind, the metaphor may have resonated even further. Whereas Hooke was a crooked giant whose shoulders could scarcely sustain Newton's weight, Descartes had been a blind, and dangerously misguided, giant. In old age, Newton conceded he had been brought up a Cartesian, only to add that Descartes' blunders set him on the right track. Both William Whiston in the mid-1690s, and Conduitt three decades later, testified that Newton attributed "his amazing Theory of Gravity" to his early cognizance of Descartes' errors. For his part, Voltaire passed on the story that, having read a few pages of Descartes' *Principia philosophiae*, Newton got tired of soiling the pages with the annotation "error" so he threw the book away. The tale may be apocryphal. Or Voltaire may have confused Newton's reaction with his response to another Cartesian text, the *Géometrie*. (That copy survives, and it is marked with a good dozen instances of "error," "I hardly approve," and "this is no geometry.")[35] What matters is that by the 1670s, Newton had become vehement in his censure of Cartesian mathematics. When investigating the Greek solid locus, he lashed out at Descartes' pretentious claim in the *Géometrie* to have "achieved something so earnestly sought after by the Ancients" when the topic was no mystery at all. The method of the ancients, Newton insisted, "is more elegant by far than the Cartesian one. For he achieved the result by an algebraic calculus which, when transposed into words (following the practice of the Ancients in their writings), would prove to be so tedious and entangled as to provoke nausea, nor might it be understood." Newton came even to eschew his own discoveries – and turn his back on analysis – because of his increased aversion toward Cartesian mathematics.[36]

Newton's repudiation of Descartes is equally evident in his natural philosophy. The Frenchman is scarcely mentioned in the *Principia*. His name occurs in passing on a few occasions and once more distinctly – in order, it seems, to deny him the discovery of Snel's law of refraction. If anything, Descartes is conspicuous by his absence. The very title page of Newton's masterpiece was designed to signal to the cognoscenti Newton's vanquishment of his rival: ***PHILOSOPHIAE*** *NATURALIS* ***PRINCIPIA*** *MATHEMATICA*. The title of Descartes' book, in boldface and set in larger type, is fully incorporated into Newton's title. (By the third edition, scarlet replaced the boldface.) The qualifiers, to my mind, convey simultaneously a message of modesty and transcendence: Newton's principles – so the unbolded lettering suggests – are neither boastful pretensions, founded on flimsy foundations of romantic physics and pernicious metaphysics, nor principles presuming to erect a new philosophy. Newton's concern is to establish natural philosophy on the secured foundations of mathematics, wholly absent in Descartes' treatise. This message was reinforced by the substitution of "Axioms, or the Laws of Motion" for Descartes' more pretentious "Certain Rules or Laws of Nature." In the text itself, the anti-Cartesian crusade is everywhere evident. Galileo is credited with the discovery of the first two laws of motion, as if to preclude any need to credit Descartes. (Newton never even hinted at any debt to the latter's concept of inertia.) Newton may have composed Book II, devoted to the resistance of bodies in fluids, purposely in order to destroy Descartes' system, as some have suggested. He certainly concluded that book by demonstrating the incompatibility of Cartesian vortices with Kepler's laws.[37]

There exists a unique manuscript among Newton's papers entitled "De gravitatione et aequipondio fluidorum" ("On the Gravity and Equilibrium of Fluids"), which is uncharacteristic both in its openly virulent attack on Descartes and in its extended focus on metaphysics. Such atypical features have led scholars to date it back to Newton's first encounter with the *Principia philosophiae* in the mid-1660s. However, its origins might be more precisely dated to around 1671, and to a course of lectures Newton delivered at Cambridge against Descartes' mechanics and Henry More's hydrostatics.[38] A decade or so later, I would argue, Newton contemplated reworking his earlier lectures into a more sustained philosophical argument against Descartes; he was sensible that the "general inclination, especially of the brisk part of the university, to use him" – as a contemporary undergraduate put it – would prove an impediment to the reception of his ideas.[39] "De gravitatione" was perhaps an attempt to confront this impediment head on. Newton controverted some key issues of Cartesian natural philosophy and metaphysics: his text disposed of the equivalency of matter and extension; it demolished, at least to Newton's satisfaction, the conception of the relativity of motion; and it dismissed the Cartesian plenum (a space completely filled with matter), as well as his "subtle matter," as an enabling medium of motion. Newton's discussion of these and related issues is informed, again unique for this period, by deep theological considerations.

If Newton intended to publish "De gravitatione," he quickly relinquished the idea. Instead, he concluded section 11 of Book I of the *Principia* by assuring his readers that its mathematical investigations held great promise for philosophy more generally:

> Mathematics requires an investigation of those quantities of forces and their proportions that follow from any conditions that may be supposed. Then, coming down to physics, these proportions must be compared with the phenomena, so that it may be found out which conditions [or laws] of forces apply to each kind of attracting bodies. And then, finally, it will be possible to argue more securely concerning the physical species, physical causes, and physical proportions of these forces.[40]

In later years, Newton was to make even more forceful statements regarding the value of his mathematical principles to the practice of philosophy, in all its parts. But for the time being, he released the *Principia* in the austere, almost forbidding, format we know today.

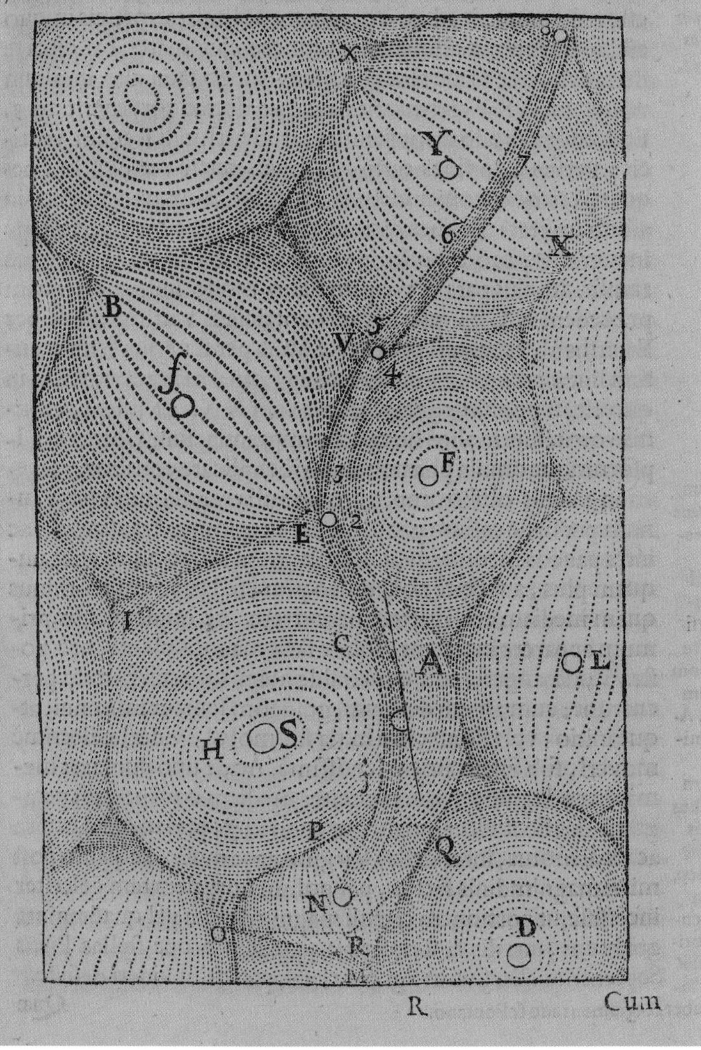

PARS TERTIA. 129

rius aſcendat ſecundum lineam curvam C 2, quæ eo minus diſtat à recta tangente, quo hoc Sidus ſolidius eſt, & quo majori cum celeritate delatum eſt ab N ad C.

R. Cum

In the Cartesian system, the sun and the planets are enveloped within large whirlpools (vortices) of subtle matter that pervades space; the swerving of these vortices produces the discernible motion of the heavenly bodies. Newton was brought up a Cartesian, later claiming that Descartes' blunders set him on the right track. The vortices are represented here in a plate from Descartes' *Principia philosophiae* (Amsterdam, 1656). – NYPL–General Research Division

2

Isaac Newton in 1689, two years after the publication of the first edition of the *Principia*. Painting by Sir Godfrey Kneller. – Courtesy of the Trustees of the Portsmouth Estate

THE LION'S CLAWS

Ironically, the *Principia,* as we have seen, may have owed its publication to the persistent efforts of Robert Hooke to elicit from Newton a solution to his question regarding the shape of an orbit given force. Frustrated by Newton's unwillingness to posit an answer, Hooke kept returning to what he perceived as both his finest accomplishment and greatest challenge: how to make the grand leap from a partly intuitive, partly experimental conception of a central force to a mechanics of orbital motion? Indeed, it was precisely his deep awareness of the enormous difficulties involved that had prompted him to seek Newton's assistance. After the Lucasian Professor turned a cold shoulder, Hooke continued to dwell on the problem with his friends the architect Sir Christopher Wren and the astronomer (and Clerk of the Royal Society) Edmond Halley. In 1684, during one of their meetings in London, Hooke probably suggested that Halley make a trip to Cambridge in

Astronomer Edmond Halley (1656–1742) cajoled Newton into writing the *Principia*, which Halley published at his own expense and then enthusiastically promoted at home and abroad. – © The Royal Society

the hope that he might have better luck in extracting a solution from the reticent professor. Certainly, the question that Halley put to Newton in August of that year was identical to the one Hooke had propounded three and a half years earlier, when he asked Newton what he "thought the Curve would be that would be described by the Planets supposing the force of attraction towards the Sun to be reciprocal to the square of their distances from it." An ellipse, Newton responded instantly. "Struck with joy and amazement," Halley asked him how he knew this. "Why saith he I have calculated it." But when Halley requested to see the calculation, Newton could not locate it among his papers, but promised to send it along to London.[1]

Three months later, Halley had in hand not only the desired demonstration, but a newly composed nine-page paper – "De motu corporum in gyrum" ("On the Motion of Bodies in Orbit") – which included much more besides. Astounded and delighted, Halley rushed back to Cambridge and persuaded Newton to once again venture publication. For eighteen months Newton became the recluse whose image has persisted for more than three centuries, working at a feverish pace on a book that took shape in the process of composition. For his part, Halley continued to cajole Newton in the broadest sense of the word: offering technical advice, correcting proofs, even bearing the cost of printing. Most important of all, he soothed Newton's ire (and threat to forgo publication) upon hearing that Hooke demanded that Newton acknowledge the assistance he had received from him. It is surely an accurate conclusion that were it not for Halley's "great zeal, able management, unwearied perseverance, scientific attainments, and disinterested generosity, the *Principia* might never have been published."[2] On July 5, 1687, Halley could finally inform Newton: "I have at length brought your Book to an end, and hope it will please you."

Halley's selfless efforts on behalf of the *Principia* caused him to neglect his duties as Clerk of the Society. In early April 1687, he intimated to the mathematician John Wallis that he had "lately been very intent upon the publication of Mr Newtons book, which has made me forget my duty in regard of the Societies correspondants; but that book when published will I presume make you a sufficient amends for this neglect."[3] Not all correspondents were neglected, however. Hand in hand with his editorial activities, Halley embarked on a propaganda campaign intended to prepare the ground for the diffusion of the gospel according to Newton on the Continent. Influential foreign correspondents of the Royal Society proved the ideal medium to spread the message. One such member was Johann Christoph Sturm, professor of mathematics and physics at Altdorf, to whom Halley sent a précis of the new theory of universal gravitation, so "brilliantly investigated" by Newton. Another was Salomon Reisel, physician to the duke of

Würtemburg, who was apprised of "a truly outstanding book" written by perhaps the greatest geometer ever to exist, someone whose *Principia* will prove "how far the human mind properly instructed can avail in seeking truth."[4] Once the *Principia* saw publication, Halley followed up his preparatory work by dispatching numerous presentation copies to these and other individuals. Gian-Domenico Cassini, director of the French Observatory, was one such recipient; Vincenzo Viviani, Galileo's last disciple, was another. Viviani was sufficiently impressed to show his copy to Grand Duke Cosimo III de Medici.

The propaganda campaign persisted in the periodical literature. Halley himself led the way with an (anonymous) pre-publication review in the *Philosophical Transactions*, which, conveniently, he also edited. He used the occasion to wax eloquent on the "incomparable" treatise that provided "a most notable instance of the extent of the powers of the Mind; and has at once shewn what are the Principles of Natural Philosophy, and so far derived from them their consequences, that he seems to have exhausted his Argument, and left little to be done by those that shall succeed him." The five pages that followed offered a lucid summary of key Newtonian contributions.[5] A more extensive, but equally appreciative, review by the Leipzig mathematician Christoph Pfautz appeared in the chief German periodical, the *Acta Eruditorum*. In eighteen quarto pages, Pfautz managed to present more than just a flavor of the purpose and achievement of Newton's magnum opus: he provided a good summary of the laws of motion; an overview of the manifold topics treated in Books I and II; a summary of the potent critique of the Cartesian vortices; and a good outline of Book III and of universal gravitation.

Pfautz's Latin review had been preceded by John Locke's French contribution to Jean LeClerc's *Bibliothèque Universelle*. Locke first read the *Principia* in September 1687 and reputedly approached Christiaan Huygens to ask "whether all the mathematical *Propositions* ... were true." Having been so reassured, Locke "took them for granted, and carefully examined the Reasonings and *Corollaries* drawn from them, became Master of all the Physics, and was fully convinced of all the great Discoveries contained in that Book." When commissioned by LeClerc a few months later, Locke read the *Principia* again. Appearing in March 1688, his review commenced with a clear statement regarding the significance of Newton's geometrization of mechanics. Locke followed with a straightforward rendering into French of the section headings of Books I and II, as well as a more detailed account of the manner in which Newton demolished Cartesian vortices, and a fuller account of Book III. Least informative, and bordering on the hostile, was the two-page review in the *Journal des Sçavans* for August 1688, probably written by the orthodox

In the second edition of his *Essay Concerning Human Understanding*, the philosopher John Locke (1632–1704) noted that Newton, "in his never enough to be admired" *Principia*, "has demonstrated several Propositions, which are so many new Truths, before unknown to the world." – National Portrait Gallery, London

Cartesian Pierre-Sylvain Régis. Beginning with the backhanded compliment that Newton's work "is a mechanics, the most perfect that one could imagine," the reviewer confesses "that one cannot regard these demonstrations otherwise than as only mechanical." Régis – assuming he was the author – proceeds to subvert Newton's meaning in definition 8 by asserting that the author himself "has not considered their Principles as a Physicist, but as a mere Mathematician." The alleged hypothetical nature of the *Principia* is stressed even more in reference to Newton's exposition of the System of the World, which, the reviewer is certain, was founded on arbitrary hypotheses that could hardly serve as a foundation for a new physics. Régis concludes with a touch of sarcasm: "In order to make an *opus* as perfect as possible, M. Newton has only to give us a Physics as exact as his Mechanics. He will give it when he substitutes true motions for those that he has supposed."[6]

These reviews, combined with Halley's propagandist efforts, assured that the publication of the *Principia* became a major event and that Newton was ensconced as a peerless mathematician and natural philosopher. And yet, in retrospect at least, the diffusion of the *Principia* appears to have been arrested for decades. Only in the middle of the eighteenth century – after decades of fierce opposition – did it ascend to that pinnacle of scientific achievement associated with it ever since. In his 1727 *éloge* of Newton, Bernard de Fontenelle, perpetual Secretary of the French Académie des Sciences, reflected on the slow reception of the *Principia*, opining that this delay could be attributed to its recondite nature: "In the end, when the book was sufficiently well known, all the applause that it had won so slowly broke out on all sides, constituting a single paean of praise. Everyone was struck by the original intelligence shining through the book; by the creative spirit which has throughout a fortunate century been shared with only three or four men among all the most learned nations." In 1727, however, total conversion was still a thing of the future; indeed, Fontenelle himself was destined to contribute one last anti-Newtonian salvo a quarter of a century later (see chapter 3 below). More on target was Jean-Baptiste Biot's assessment on the eve of the first centenary of Newton's death: "More than fifty years elapsed before the great physical truth contained and demonstrated in the *Principia* was, we do not say followed up and developed, but even *understood* by the generality of learned men."[7]

The difficulties encountered by prospective readers of the *Principia* were considerable. Humphrey Babington, Newton's old college benefactor, is purported to have said he "might study seven years" before he "understood anything of it." The reaction of the venerable Doctor of Theology was undoubtedly typical of the sinking feeling of an older generation. Consider the sixty-one-year-old mathematician and (Cartesian) natural philosopher Gilbert Clerke, who, within two months of publication, wrote Newton: "I confesses I doe not as yet well understand so much as your first three sections, for which you doe not require that a man should be *mathematice doctus*." Clerke went on to apologize profusely that, having spent his declining years in obscure retirement where he was unfamiliar with the "brave notions" of the likes of Galileo and Huygens, he all but "despaire[d] of understanding" the *Principia*. Still, he desired to learn about the tides and other natural phenomena, and he hoped that Newton would forgive his importunities as he [Clerke], together with Isaac Barrow, had done more than anybody else to introduce such studies into Cambridge four decades earlier.[8]

Nor was this sort of struggle with the *Principia* symptomatic only of the old guard. Younger aspirants, too, often felt out of their depth. The mathematician Colin Campbell, two years younger than Newton, was fired up

The natural philosophy of René Descartes (1596–1650) was gradually eclipsed by Newton's in the course of the eighteenth century. – NYPL–Print Collection

to study the book in late 1687 by University of Edinburgh mathematics professor David Gregory, who also told him that "Newton will take you up the first month you have him." That turned out to be a generous estimate. Campbell borrowed a copy from Gregory's student John Craig, who by then had mastered barely a quarter of the book, and found even the preparation of an account of the *Principia* "to be a task of no small trouble"; a year later, when Craig requested the book's return as another friend wished to borrow it, Campbell complained that it still gave him great trouble. And in London, the twenty-one-year-old teacher of mathematics Abraham de Moivre encountered the *Principia* at the home of the Duke of Devonshire. He

> opened the book and deceived by its apparent simplicity persuaded himself that he was going to understand it without difficulty. But he was surprised to find it beyond the range of his knowledge and to see himself obliged to admit that what he had taken for mathematics was merely the beginning of a long and difficult course that he had yet to undertake. He purchased the book, however; and since the lessons he had to give forced him to travel about continuously, he tore out the pages in order to carry them in his pocket and to study them during his free time.[9]

Such difficulties were shared by numerous mathematicians and natural philosophers. But the checkered career of the *Principia* cannot be attributed solely to the considerable effort involved in mastering it. The truth is that by the end of the seventeenth century, several factors had combined to create an antagonistic environment that proved detrimental to the reception of the *Principia* as well as of the *Opticks*. Such antagonism was not inevitable, however, notwithstanding the challenge that Newton's teachings posed to the most cherished assumptions held by the vanguard of the scientific and philosophical community. A scrutiny of the attempts made by two other contemporary titans, Christiaan Huygens and Gottfried Wilhelm Leibniz, to come to terms with Newton's masterpiece yields some interesting insights about what the reception of the *Principia* might have been under *different* circumstances.

Huygens was perhaps the only seventeenth-century savant to share with Newton both mathematical genius and a keen grasp of physical theory tethered to uncanny experimental skills. Also like Newton, Huygens was profoundly disillusioned with Descartes. As he recounted in 1693:

Mr. Descartes had found the way to have his conjectures and fictions taken for truths. And to those who read his Principles of Philosophy something happened like that which happens to those who read novels which please and make the same impression as true stories. The novelty of the images of his little particles and vortices are most agreeable. When I read the book of Principles the first time, it seemed to me that everything proceeded perfectly; and when I found some difficulty, I believed it was my fault in not fully understanding his thought. I was only fifteen or sixteen years old. But since then, having discovered in it from time to time things that are obviously false and others that are very improbable, I have rid myself entirely of the prepossession I had conceived, and I now find almost nothing in all his physics that I can accept as true, nor in his metaphysics and his meteorology.[10]

The Dutch savant Christiaan Huygens (1629–1695) shared with Newton both mathematical genius and uncanny experimental skills, as well as a profound disillusionment with Descartes. – Collection Haags Historisch Museum, The Hague

Huygens, however, was recalcitrant about substituting Newton's principle of attraction for Descartes' errors. Before even seeing the *Principia*, he confided to the Swiss mathematician Nicolas Fatio de Duillier that he was untroubled by Newton's rejection of Descartes, "provided he does not give us suppositions like that of attraction." A few months later, now with a presentation copy from Newton in hand, he stood in awe. The "famous M. Newton," he jotted down in mid-December 1687, "has brushed aside all the difficulties [concerning the Keplerian laws] together with the Cartesian vortices; he has shown that the planets are retained in their orbits by their gravitation toward the Sun. And that the excentrics necessarily become elliptical." A fortnight later, he even wrote his brother Constantijn, then in England, that he wished he too could be there, "only to make the acquaintance of Mr. Newton whom I exceedingly admire for the beautiful inventions that I found" in the *Principia*.

Huygens's admiration was clearly sincere; so, too, must have been the painful awareness that he had come so close to preempting Newton – and yet the feat had eluded him. Conversion, however, was not an option he entertained. When he learned that the *Principia* was going through press, Huygens was strengthened in his resolve to rewrite a lecture he had delivered to the French Académie des Sciences in 1669, in which he grafted a theory of terrestrial gravity onto Descartes' vortices. The *Discours de la cause de la pesanteur* ("Discourse on the Cause of Gravity") appeared in 1690, with an appendix commenting on Newton's theory. There, despite rejecting much of Descartes' physics – including the identification of matter and extension, and the existence of plenum – Huygens reaffirmed his unwavering allegiance to contact as the sole explanatory principle of the mechanical philosophy. He told Newton as much in person when he visited London during the summer of 1689; to Leibniz, a year later, he intimated that he regarded as "absurd" all the theories that Newton had established upon the principle of attraction.[11] Unlike so many readers of Newton, however, Huygens grasped full well that Newton had established his principle of attraction on *empirical* foundations and that to controvert the *Principia* one needed to marshal empirical, not philosophical, proofs. Such a realization – combined with a commitment to take empirical evidence seriously – might have swayed Huygens to change his mind about universal gravitation in due course. But in the last few years of his life (he died in 1695), Huygens was not entirely persuaded that Newton himself was committed to action at a distance. The source of this sense of a possible ambiguity in Newton's position was Fatio de Duillier, who proclaimed himself Newton's mouthpiece and official interpreter.

Fatio, born in Geneva in 1664, was educated under Jean-Robert Chouet at the local academy. Following a brief sojourn in Holland, he arrived in England in June 1687, carrying in him the belief that the two greatest "philosophers in the world" were Robert Boyle, "for the detail of his experiments concerning our earthly bodies," and Christiaan Huygens, "for physics in general, above all in those areas in which it is involved with mathematics." Active campaigning secured him a fellowship at the Royal Society in May 1688; within a month, he wrote Huygens that his new colleagues "have reproached me that I was too much of a Cartesian and had given me to understand that since the meditations of [Newton] all of Physics was very much changed." Within a year and a half, he was converted. Newton, he wrote his old teacher Chouet – who had been instrumental in introducing Cartesianism to Geneva two decades earlier – was the most cultivated man (*le plus honnête homme*) he knew "and the greatest mathematician who has ever lived." He "discovered geometrically the true System of the World in a way that leaves no doubt in the minds of those who are able to understand it." Fatio continued:

> it's all over for vortices ... which were only an empty imagination. The whole system of Descartes and his whole world, so filled that it was impossible to turn around in it, are no more than reveries that one ought to take pleasure in laughing at after he is instructed in the truth. Nothing has ever been discovered in so grand, so noble and so complete a manner as that which Monsieur Newton has shown us.[12]

The Swiss mathematician Nicolas Fatio de Duillier (1664–1753) was an enthusiastic but overzealous convert to Newtonianism, who exaggerated his influence on Newton's ideas. – Genève, Bibliothèque publique et universitaire, Collections iconographiques

By this time, Newton was also completely taken with the young Swiss, so much so that Fatio developed a strong sense of entitlement, with confidence bordering on bravado. Fatio repeatedly assured Huygens that Newton would prove receptive to whatever the Dutchman proposed. "So many times I have found him ready to amend his book upon matters of which I spoke to him," Fatio wrote Huygens in February 1690, "that I cannot sufficiently marvel at his facility, particularly as to those matters which you have criticised."[13] Fatio was also convinced that Newton would entrust him with editing a new edition of the *Principia* – and broadly advertised his confidence. Unfortunately, insofar as the reception of the *Principia* was concerned, Fatio also insinuated that Newton was quite receptive to his own mechanical explanation of the cause of gravity. Fatio presented a preliminary version of his theory – which, he said, had been received favorably by Huygens – at a meeting of the Royal Society on July 4, 1688. He continued to work on it, informing his brother in June 1690 that Newton and Halley believed it was true. Indeed, Fatio, who was in the habit of asking those to whom he conveyed his theory to sign the manuscript, obtained Newton's and Halley's signatures on March 19, 1690. (Huygens signed the manuscript the following year, Leibniz in 1694, and Jakob Bernoulli in 1701.) Fatio later added that Newton inserted his favorable opinion into his own copy of the first edition of the *Principia* when preparing the second edition: "There he did not scruple to say *That there is but one possible Mechanical cause of Gravity, to wit that which I had found out*. Thô he would often seem to incline to think that Gravity had its Foundation only in the arbitrary Will of God." Fatio's confidence knew no bounds, and it must have carried considerable weight with contemporaries. David Gregory, for example, recorded on December 28, 1691, that Fatio "designs a new edition" of the *Principia*, "wherein among a great many notes and elucidations, in the preface he will explain gravity acting as Mr Newton shews it doth, from the rectilinear motion of particles the aggregate all which is but a given quantity of matter dispersed in a given space. He says that he hath satisfied Mr Newton, Mr Hugens and Mr Hally in it."[14]

A few years later, Gregory supplemented the entry: "Mr Newton and Mr Hally laugh at Mr Fatios manner of explaining gravity." The change is indicative of Newton's abrupt termination, in the summer of 1693, of his relationship with Fatio – an event that contributed to Newton's mental breakdown in 1693–94. But by then the damage was done; among influential Continental savants, Newton's commitment to a non-mechanical cause of universal gravitation was thought to be tentative, and he might prove receptive to corrections. Indeed, a purposeful reading of a letter Newton sent Leibniz in late 1693 could suggest such an interpretation. Responding to a friendly letter from Leibniz, in which the German philosopher expressed his admiration for the *Principia* (although, he wrote, he was inclined to believe that gravity is "caused or regulated by the motion of a fluid medium, on the analogy of gravity and magnetism as we know it here"), Newton was a paragon of courtesy. He, too, valued their friendship, and was always happy to explain the reasons behind his discoveries; if "someone explains gravity along with all its laws by the action of some subtle matter, and shows that the motion of planets and comets will not be disturbed by this matter, I shall be far from objecting."[15] Two decades later, in the new "General Scholium" he added to the second edition of the *Principia* (1713), Newton remained famously noncommittal: "I have not as yet been able to deduce from phenomena the reason for these properties of gravity, and I do not feign hypotheses." It was enough for him that "gravity really exists and acts according to the laws" he had established.[16]

The German philosopher and mathematician Gottfried Wilhelm Leibniz (1646–1716) was for years locked in a bitter priority dispute with Newton over which of them deserved credit for the invention of the calculus.
– NYPL–Print Collection

By 1713, the stakes had become much higher, and Leibniz contributed to making them so. However, in the decade and a half following the publication of the first edition of the *Principia*, Leibniz was keen to maintain an open, constructive, and friendly attitude toward Newton. Like Huygens, he held Newton in very high esteem; and again like Huygens, he was affected deeply by the reading of the *Principia*, so much so that Leibniz succumbed, quite uncharacteristically, to the temptation of launching a deceitful publication. While in Vienna, during the second half of 1688, he received a copy of the *Principia* that Newton had sent him. Hastily, Leibniz composed two brief treatises – *Tentamen de motuum coelestium causis* ("An Essay upon the Causes of the Celestial Motions") and *Schediasma de resistentia medii et motu projectorum* ("Papers on the Resistance of the Medium and on the Motion of Heavy Projectiles in a Resisting Medium") – and rushed them into print in the *Acta Eruditorum*. In his cover letter to the editor, Otto Mencke, Leibniz claimed that his travels had kept him out of touch with new books and, consequently, he had not yet seen the *Principia*. However, the chance reading of Pfautz's review of that "celebrated" author in the *Acta* had prompted him to publish his own thoughts on these matters, "so that the sparks of truth should be struck out by the clash and sifting of arguments, and that we should have the penetration of a very talented man [Newton] to assist us."[17]

Nothing was particularly wrong with Leibniz's attempt to stake his priority on certain issues made famous with the publication of the *Principia*. Nor was it reprehensible of him to attempt a refutation. Huygens, as noted above, had done just that when rewriting his thoughts on gravity; Hooke had made one final heroic effort to crack the problem of orbital motion for the same reason. But whereas Huygens was careful to acknowledge that his additions to his *Discours de la cause de la pesanteur* were written with full knowledge of Newton's *Principia*, Leibniz lied about his knowledge of the book and its contribution to the formulation of his diverging ideas on the matter. Contemporaries may have doubted his sincerity – as Newton would years later – but Leibniz stuck to his story. Only recently has the evidence regarding his duplicity come to light.

Such human foibles aside, Leibniz was not shy about expressing his admiration for Newton. This "remarkable man," he wrote Mencke in that infamous cover letter, "is one of the few who have advanced the frontiers of the sciences." When informed by Huygens of Newton's recovery from his breakdown in

1694, Leibniz expressed joy, adding that he wished longevity and health to Huygens and Newton "in preference to others, whose loss would not be great, speaking comparatively." Finally, in 1701, when asked by the Prussian Queen for his opinion of Newton, Leibniz "said that taking Mathematicks from the beginning of the world to the time of Sir I[saac] What he had done was much the better half."[18] We must accept the sincerity of Leibniz's sentiments. His profound disagreement with Newton on certain issues notwithstanding, Leibniz made good on his word that what really concerned him was the common pursuit of truth. Writing to Newton in 1693, Leibniz reiterated the message he had conveyed to Mencke in 1688: "it is by the friendly collaboration of you eminent specialists in this field that the truth can best be unearthed."[19] To this end, he also implored Newton – repeatedly – to publish his results. He also attempted to persuade Newton's friends to do the same, cognizant of how reluctant Newton was to appear in print. The Royal Society, he wrote James Brydges in October 1694, should pressure the Professor, recently recovered from his breakdown, to "publish his further thoughts and improvements on the subject of his late book," together with his other mathematical and physical discoveries, "lest by his death they should happen to be lost." Two years later, upon hearing of Newton's appointment as Warden of the Mint, Leibniz expressed deep regret that such an honor would divert Newton from his intellectual pursuits.[20]

While his esteem for Newton rivaled that of Huygens, Leibniz, with all his mathematical genius, was far more of a philosopher in the rationalist tradition than was the Dutchman. To be sure, Leibniz was quite critical of Cartesian philosophy – and not only of Descartes' physics – but his disillusionment was never as profound as the disillusionment of either Newton or Huygens. In fact, Leibniz's criticism of Descartes was informed by similar philosophical convictions – for example, on the matter of final causes. Hence, whether he argued with Descartes or with Newton, the metaphysical commitments of Leibniz determined his conception of natural philosophy. It is hardly surprising, then, that Leibniz did not share the conviction of his two illustrious contemporaries regarding the primacy of empirical evidence in natural philosophy. Nor was he ever free to devote himself to the demanding and time-consuming calculations that so distinguished his Dutch mentor and English rival. Partly for this reason, by the early years of the eighteenth century, Leibniz was forced to abandon the physical program he had outlined in the *Tentamen* to rival the program propounded in the *Principia*.

Back in 1688, Leibniz had made precisely such an effort when he realized that a "purely philosophical response would have left Newton as the only master of celestial mechanics."[21] This perception of Leibniz proves Newton right in his resolution to restrict the scope of the *Principia*. Indeed, were it not for the divisiveness that swept aside good will and manners, Leibniz's effort to controvert Newton's "poor" philosophy would have focused, quite likely, less on metaphysics and more on natural philosophy. Furthermore, a more propitious atmosphere would have ushered in the promise that the significant progress made by Leibniz's disciples in converting the *Principia* into the language of the calculus would go beyond the domain of pure mathematics; likewise, that the mathematicians Johann and Jakob Bernoulli and the cadre they helped raise would also devote themselves to issues of natural philosophy so central to the Newtonians. Unfortunately, the eruption of the calculus priority disputes retarded such a process for at least two decades, hardening positions and prejudices instead.

Ironically, it was a compliment of sorts that Leibniz paid Newton that sparked the celebrated priority dispute. In late 1696, Johann Bernoulli issued the brachistochrone challenge problem – to determine the curve (not on the same vertical line) along which "a heavy body

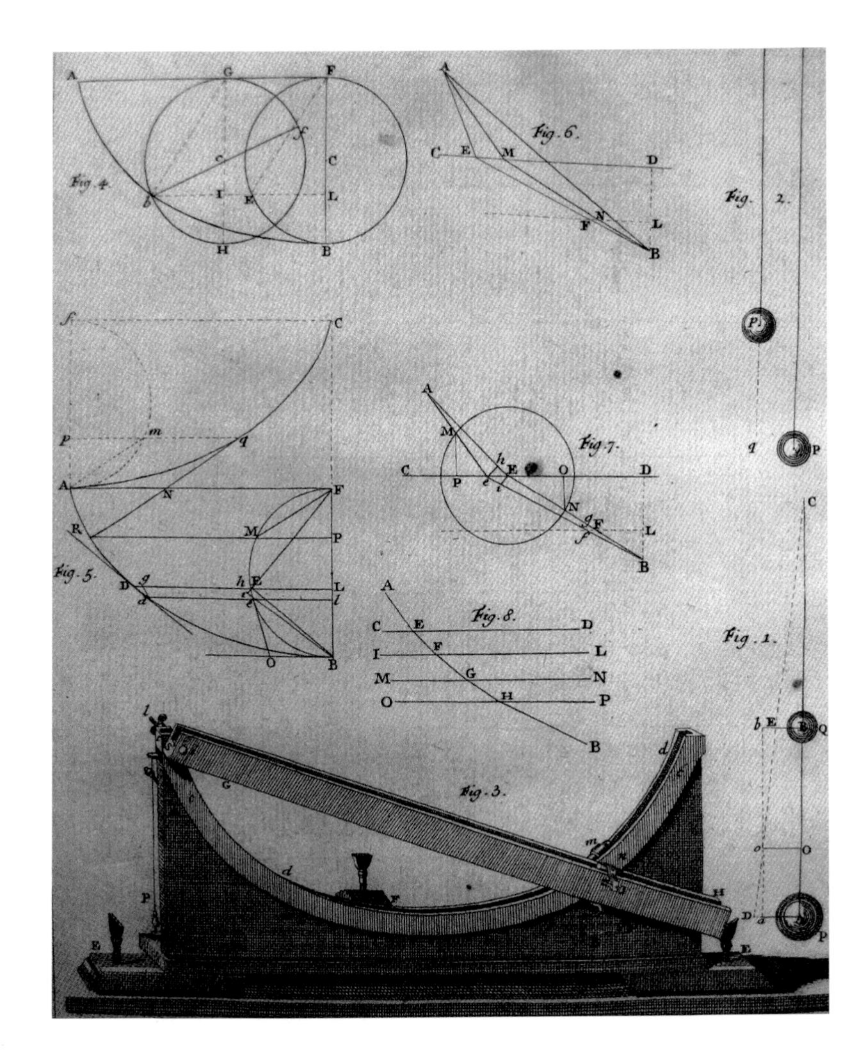

The solution to the brachistochrone challenge problem – that the shortest path along which a body will fall when not in the same vertical plane is a curve (cycloid) and not a straight line – was demonstrated through the experimental apparatus used by 's Gravesande, shown here at the bottom of the plate. Balls were dropped simultaneously along the straight and curved tracks, and the greater steepness of the latter made the ball accelerate faster. – Courtesy of the California Institute of Technology Archives

shall, under the force of its own weight, most swiftly descend from any given point A to any given point B."[22] By Newton's own account, he received the problem at four o'clock on the afternoon of January 29, 1697, after a strenuous day at the Mint – but he immediately sat down and did not go to bed until he had solved it, some twelve hours later. The solution was published anonymously in the *Philosophical Transactions*; when Bernoulli received it, he immediately recognized its author, "*tanquam ex ungue leonem*" – "as the lion is recognized by his claws." In an article published in 1697 in the *Acta Eruditorum*, Leibniz remarked that only a handful of mathematicians, who "mastered our calculus," were capable of solving such a problem: himself, the Bernoulli brothers, the Marquis de l'Hôpital, Newton, and, had he lived, Huygens.[23]

Leibniz's article infuriated Fatio, who evidently took umbrage at being excluded from the ranks of the top European mathematicians (he was not even sent the challenge). Incensed, he published a small pamphlet in 1699, wherein he claimed that as far back as 1687, and independently of Leibniz, he had developed a method suited perfectly to solving the brachistochrone problem. Fatio further availed himself of Leibniz's unfortunate turn of phrase "our calculus" in order to charge that it was actually Newton who had discovered the calculus, and that Leibniz may well have helped himself to Newton's results. Leibniz and Bernoulli clearly recognized in Fatio's

Two artistic renditions of Newton's renowned prismatic experiments: "Newton Investigating Light," from *The Illustrated London News*, June 4, 1870. - Burndy Library; and (below) "Newton and the Prism," from William Hayley, *The Life of George Romney, Esq* (London, 1809). - NYPL-Print Collection

harangue the sour grapes of a mentally unstable person; after handily dispatching him in the pages of the *Acta*, they let the matter rest. But Fatio, regardless of his mental state, probably had more on his mind than simply setting straight the historical record; he sought to restore himself in Newton's graces. However, if Fatio assumed the role of Newton's public defender in order to ingratiate himself with Newton, he was sorely disappointed. Hitherto Newton seems to have entertained no thought of staking a claim to the invention of the calculus, and the rupture with Fatio was far too deep to mend. Nevertheless, the idea, once planted in Newton's mind, slowly took shape.[24]

In 1704, Newton finally published his *Opticks*. Tradition has it that he purposely withheld publication for three decades, until after Hooke's death (in 1703). Undoubtedly, this perception was inspired by Newton's claim in the preface: though prepared in large part much earlier, he wrote, "To avoid being engaged in Disputes about these Matters, I have hitherto delayed the printing, and should still have delayed it, had not the Importunity of Friends prevailed upon me."[25] Not surprisingly, those familiar with Hooke's critique of Newton's theory of light and colors eagerly deduced, from the proximity of the publication of the *Opticks* to Hooke's death, that it was Hooke whom Newton had in mind. And yet, though Hooke remained an irritant to Newton, Newton did not withhold publication for fear of what Hooke might do. Newton may well have found it convenient in 1704 to blame unnamed cavilers for the long delay in publishing, but the truth is that the book did not exist in its present form before the turn of the eighteenth century – although parts of it had been ready for some time.

The *Opticks* was destined to be greeted as every bit as revolutionary and challenging, and every bit as controversial, as the *Principia*, notwithstanding that the radical ideas set forth – concerning the unequal refrangibility of light rays, and the heterogeneity of sunlight – had already been published by Newton, in the early 1670s. Significantly, although Newton's theories were accepted by many British practitioners by the time the *Opticks* saw publication in 1704, Newton's 1672 paper appears to have faded from memory on the Continent. This occurred in part because Continental savants read English, if at all, only with great difficulty, and in part because Edmé Mariotte's failure in 1681 to replicate Newton's refraction experiments appeared to disprove the immutability of color argued by Newton. Be that as it may, the publication of the *Opticks* in early 1704 quickly became an event. "My Design in this Book," Newton began, "is not to explain the Properties of Light by Hypotheses, but to propose and prove them by Reason and Experiments." Accordingly, he deployed a methodology he had utilized previously in the *Principia*, with definitions followed by propositions that were immediately demonstrated experimentally or mathematically. In terms of subject matter, to his greatly expanded treatment of his previously published theory of light and colors, he now added discussions of such new topics as diffraction and thin plates.

The *Opticks* also included two appendices, wholly unrelated to the subject matter of the book: "A Treatise on the Quadrature of Curves" and "Enumeration of Lines of the

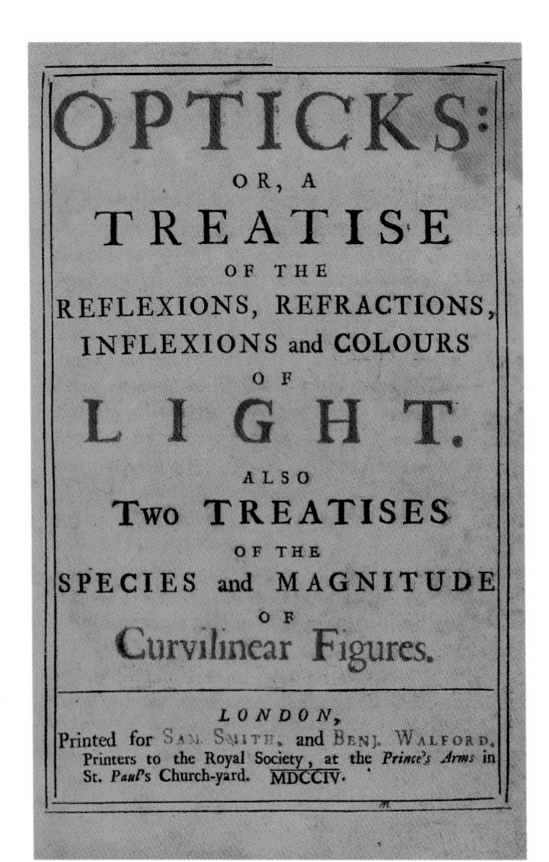
OPTICKS:
OR, A
TREATISE
OF THE
REFLEXIONS, REFRACTIONS,
INFLEXIONS and COLOURS
OF
LIGHT.
ALSO
Two TREATISES
OF THE
SPECIES and MAGNITUDE
OF
Curvilinear Figures.

LONDON,
Printed for SAM. SMITH, and BENJ. WALFORD,
Printers to the Royal Society, at the *Prince's Arms* in
St. *Paul's* Church-yard. MDCCIV.

Newton's *Opticks: or, A Treatise of the Reflexions, Refractions, Inflexions and Colours of Light* (London, 1704) was greeted as every bit as revolutionary, challenging, and controversial as the *Principia*, although the radical ideas set forth – concerning the unequal refrangibility of light rays and the heterogeneity of sunlight – had already been published by Newton, in the early 1670s. – NYPL-SIBL

Third Order." Newton was provoked into printing these old efforts, he told his friends, by the publication the previous year of George Cheyne's *The Inverse Method of Fluxions*. As he put it in the "Advertisement" to the *Opticks*, he had lent out a manuscript that contained results he had discovered decades earlier and, having found some of them printed by another hand, he decided to publish them himself. There is no reason to doubt the truth of Newton's account. Viewed from the Continent, however, the near simultaneous publication of Cheyne's book – which refused to acknowledge any results not obtained by British mathematicians – and Newton's publication of significant, but (by then) hardly novel, mathematical treatises from the mid-1660s, seemed ominous.
As far as the *Opticks* was concerned, Leibniz, reading it shortly after publication, found it "profound." He continued to champion Newton's optical theories for another decade, and so did Bernoulli. As for the two mathematical treatises, Leibniz regarded them as respectable productions, but ones that offered little that was new. His review, however, included a sentence that would later be interpreted as a devious charge that Newton was indebted to Leibniz: "instead of the Leibnizian differences Mr Newton employs, and has always employed, fluxions."[26]

By now, the Newtonian universe was slowly closing in on Leibniz, moving alarmingly beyond higher mathematics and celestial mechanics into the realm of philosophy itself. In 1702, there appeared the first two distinctly Newtonian textbooks, John Keill's *Introductio ad veram physicam* ("Introduction to the True Physics") and David Gregory's *Astronomiæ, physicæ et geometriæ elementa* ("Elements of Astronomy, Physical and Geometrical"). The former book represented the content of Keill's Oxford lectures, which inaugurated the practice of rendering Newtonianism more palatable (and intelligible) through experiments. As its title makes clear, Keill was explicit in his denunciation of Cartesian mechanism as well as Cartesian reasoning, which he exposed as conducive to all sorts of errors. In contrast, he declared, the combined effort of all the mechanical philosophers that ever existed "does not amount to the tenth part of those Things, which Sir Isaac Newton alone, through his vast Skill in Geometry, has found out by his own sagacity." Gregory did for the heavens what Keill did for the terrestrial world: he rendered Newtonian celestial mechanics teachable. However, he also used the opportunity to (discreetly) criticize Leibniz's *Tentamen*, pointing out, among other things, the incompatibility of Leibniz's cherished harmonic vortex theory with either Kepler's laws or with the theory of comets.[27] Leibniz read the favorable reviews of Keill and Gregory in the *Acta*; he also read Gregory's book and penned a not altogether successful rejoinder.

Even more significant for Leibniz's evolving conception of Newtonianism, however, was his reading (also in 1702) of Pierre Coste's French translation of John Locke's *An Essay Concerning Human Understanding*. A decade earlier, Leibniz had read the English version, but owing to his inadequate grasp of the language, he had failed to appreciate many of Locke's arguments. Now, recognizing the full challenge the Englishman posed to his philosophy, Leibniz set to writing his

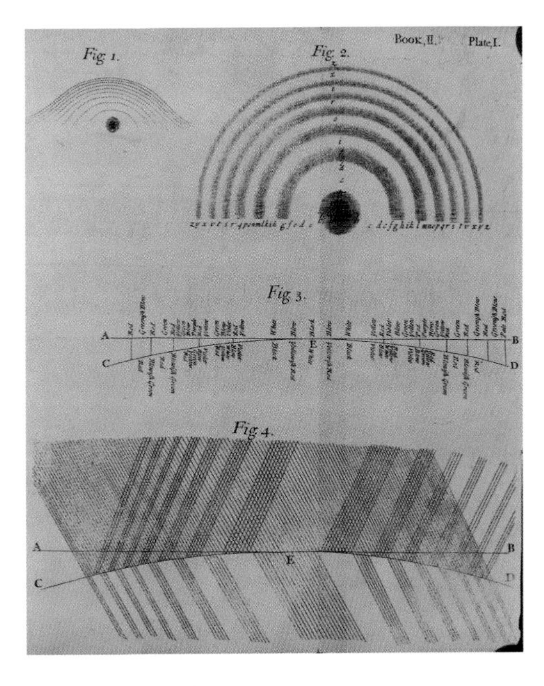

A diagram from Newton's *Opticks* (1704), – NYPL–SIBL

Nouveaux essais. He completed it in 1703, but withheld publication because of Locke's death in 1704. Leibniz surely had not missed Locke's positioning himself squarely within the Newtonian orbit. "The Commonwealth of Learning," Locke announced in his "Epistle to the Reader,"

> is not at this time without Master-Builders, whose mighty Designs, in advancing the Sciences, will leave lasting Monuments to the Admiration of Posterity; But every one must not hope to be a *Boyle*, or a *Sydenham*; and in an Age that produces such Masters, as the Great *Huygenius*; and the incomparable Mr. *Newton*, with some other of that strain; 'tis Ambition enough to be employed as an Under-Labourer

in clearing Ground a little, and removing some of the Rubbish, that lies in the way to Knowledge.

To the second edition, Locke added that Newton, "in his never enough to be admired Book, has demonstrated several Propositions, which are so many new Truths, before unknown to the World, and are farther Advances in Mathematical Knowledge."[28]

Locke had been deeply impressed by the *Principia*, and not only because he found in its pages a validation of certain of his own, similar, ideas. Newton's success in applying many of these shared principles deeply impressed Locke, reinforcing his awareness of the immense potential impact of the *Principia* on the domain of science and the republic of letters at large. Equally strong was Locke's realization that readers of his own book – which, though written years earlier, was published two and a half years after the *Principia* – would also recognize the similarities in their respective methodologies. Hence, the flattering mention of Newton at the outset of the *Essay* may have been a conscious effort on Locke's part to tether his star to Newton's, even if this meant sacrificing some of his own originality.

It is hardly surprising, then, that Leibniz forged in his mind an inseparable link between Locke and Newton, which would be germane in shaping his later theological and metaphysical critique of the former. Also possible is that Locke's reluctance to enter into a philosophical correspondence with Leibniz – as well as his disparaging remarks about the German philosopher – helped shape the ground upon which the Leibniz-Clarke controversy over the religious and metaphysical implications of Newtonian science would ultimately take root. Leibniz set off the controversy in a 1715 letter to Princess Caroline, in which he explicitly charged Locke (who was now deceased) and the Newtonians with contributing to the decay of religion in England. At Princess Caroline's request, Samuel Clarke responded to Leibniz's letter, and the ensuing exchange – which amounted to five letters apiece – was published by Clarke in 1717 as *A Collection of Papers, which passed between the late learned Mr. Leibnitz and Dr. Clarke in the years 1715 and 1716: relating to the principles of natural philosophy and religion.*[29]

The calculus priority fracas officially commenced in 1710, when John Keill published a *Philosophical Transactions* paper on central forces into which he inserted, quite unexpectedly, a charge of plagiarism against Leibniz. After discussing the inverse method of tangents, he continued:

All these things follow from the nowadays highly celebrated arithmetic of fluxions, which Mr Newton beyond any shadow of doubt first discovered, as any one reading his letters published by Wallis will readily ascertain, and yet the same arithmetic was afterwards published by Mr Leibniz in the *Acta Eruditorum* having changed the name and the symbolism.[30]

Newton was undoubtedly an accomplice in this allegation; the loyal Keill would not have dared to publish such calumny without first obtaining Newton's consent. The approval of the Royal Society Council, too, was required for all publications in the *Transactions*. And who would have officiated over such a meeting? Sir Isaac Newton, who had been elected President of the Royal Society in 1703.

Rather than settle scores through the *Acta Eruditorum*, as he had previously done with Fatio, the infuriated Leibniz opted for a more honorable route: he demanded satisfaction from the Royal Society. Leibniz reasoned that, being a Fellow himself, the Society was obligated to defend him against an unprovoked slander by another Fellow. With this idea in mind, Leibniz formally addressed Secretary Hans Sloane on February 21, 1711: "Some time

Crane Court, the home of the Royal Society, London, over which Newton presided, and from which Leibniz sought redress in the priority dispute with Newton. – © The Royal Society ago Nicholas Fatio de Duillier attacked me in a published paper for having attributed to myself another's discovery. I taught him to know better in the *Acta Eruditorum*." At the time, Leibniz continued, Sloane had assured him that Newton himself "disapproved of this misplaced zeal of certain persons on behalf of your nation and himself." Yet now, another "has seen fit to renew this most impertinent accusation." Keill might not be a "slanderer"; perhaps "he is to be blamed rather for hastiness of judgement than for malice." Nevertheless, Leibniz could not

> but take that accusation which is injurious to myself as slander. And because it is to be feared that it may be frequently repeated by impudent or dishonest people I am driven to seek a remedy from your distinguished Royal Society. For I think you yourself will judge it equitable that Mr. Keill should testify publicly that he did not mean to charge me with that which his words seem to imply, as though I had found out something invented by another person and claimed it as my own.[31]

In retrospect, Leibniz's decision to seek redress from the Royal Society is inexplicably naive. He was obviously aware that Newton was President of the Royal Society; but in his indignation, Leibniz probably deceived himself into believing that his relationship with Newton over the past two decades was such that the President, and the Society over which he presided, would conduct themselves impartially. One also detects a certain deference toward Newton, which manifests itself in Leibniz's reluctance to entertain the notion that the latter might actually be guilty of shoddy practices.[32] He exculpated Newton, at least initially, from the sins committed by his over-enthusiastic epigones. And he lived to regret it.

Matters in England swiftly went from bad to worse. Called upon to explain himself, Keill immediately drafted a response to Leibniz's complaint, part of which he read at a meeting of the Royal Society on March 22, 1711. At another meeting of the Society, two weeks later, he further justified his conduct by referring to attacks on Newton in the pages of the *Acta Eruditorum*. The Society promptly instructed Keill "to draw up a true statement of the issues in dispute and also to vindicate himself from the charges of 'reflecting' particularly on Leibniz." The acquiescence of the Society was undoubtedly obtained through the encouragement of Newton, whom Keill actively prejudiced against Leibniz and his German friends. The chance survival of one letter reveals Keill citing certain critical remarks by the reviewers of the *Acta Eruditorum* about Keill's and John Freind's works, so that Newton "may gather how unfairly they deal with you." Newton was fully persuaded, and he told Sloane as much. He even drafted for Sloane the kind of curt acknowledgment he wished Sloane to transmit to Leibniz – and which also served as a cover letter to Keill's response. Newton, it is clear, was fully committed to war.

So was Keill. He has been called Newton's "dog of war"; "pitbull" might better suggest the conduct of this particularly tenacious and dangerous opponent. His rejoinder was anything but propitiatory. He did not charge Leibniz with plagiarism in any formal manner, Keill argued; he merely stated the facts. But what had been a brief digression in the original paper was now expanded to a full-blown exegesis, over 5,000 words in length. In it, Keill assembled all of the published details regarding Newton's discoveries in order to prove that he had discovered his "method of fluxions" eighteen years before Leibniz's 1684 publication in the *Acta Eruditorum*. As he proceeded, the leitmotif of plagiarism reemerged under the guise of creative theory. Referring to the two letters that Newton sent Leibniz (via Henry Oldenburg) in 1676, Keill intimated that "the hints and examples" therein contained "were sufficiently understood by Leibniz." "Aided by these hints and these examples," Keill reasoned, "even an ordinary intellect would see quite through the Newtonian method, so is it not proper to suspect that it could not be concealed from Leibniz's very sharp mind?"[33]

The conviction that a "hint" is all that a well-prepared mind requires becomes central to the dispute. Paraphrasing Descartes' "clear and distinct ideas," Keill charged that Newton furnished "such clear and obvious hints of his method to Leibniz that it was easy for the latter to hit upon the same method." The crux of the matter is made clear: "It now remains for us to inquire what were the hints that Leibniz received from Newton, whence it would be easy for him to extract the differential calculus." The guiding principle for Keill – as well as for Newton, who had all but penned the rejoinder for Keill – was that "second inventors have no rights"; Newton's reputation as the inventor of the calculus demanded the demise of Leibniz's reputation.[34]

The sort of "hint" that Leibniz's "very sharp mind" was supposed to have latched onto was the celebrated anagram Newton included in his October 1676 letter to Leibniz:

6accdae13eff7i319n4o4qrr4s8t12ux

The anagram concealed Newton's formulation of the fundamental theorem of the calculus: "Data aequatione quotcunque fluentes quantitates involvente, fluxiones invenire; et vice versa" ("given an equation involving any number of fluent quantities to find the fluxions, and vice versa"). Leibniz may well have grasped what the anagram concealed, not necessarily because of any special cryptographic skills but because he was privy to Newton's early manuscript *De analysi*.[35] Be that as it may, Keill made a forceful case, packaged within an ominous statement: Since Leibniz "possessed so many unchallengeable riches of his own, I fail to see why he wishes to load himself with spoils stolen from others."

Leibniz waited six months before indignantly replying to Sloane on December 18: "What Mr. John Keill wrote to you recently attacks my sincerity more openly than [he did] before; no fair-minded or sensible person will think it right that I, at my age and with such a full testimony of my life, should state an apologetic case for it, appearing like a suitor before a court of law."[36] After this letter arrived in London, Newton took matters into his own hands. By now he had convinced himself that it was he who was the aggrieved party, owing to Leibniz's furtive borrowing as well as to the calumnies that the *Acta* reviewers hurled against Newton and his disciples. This convic-

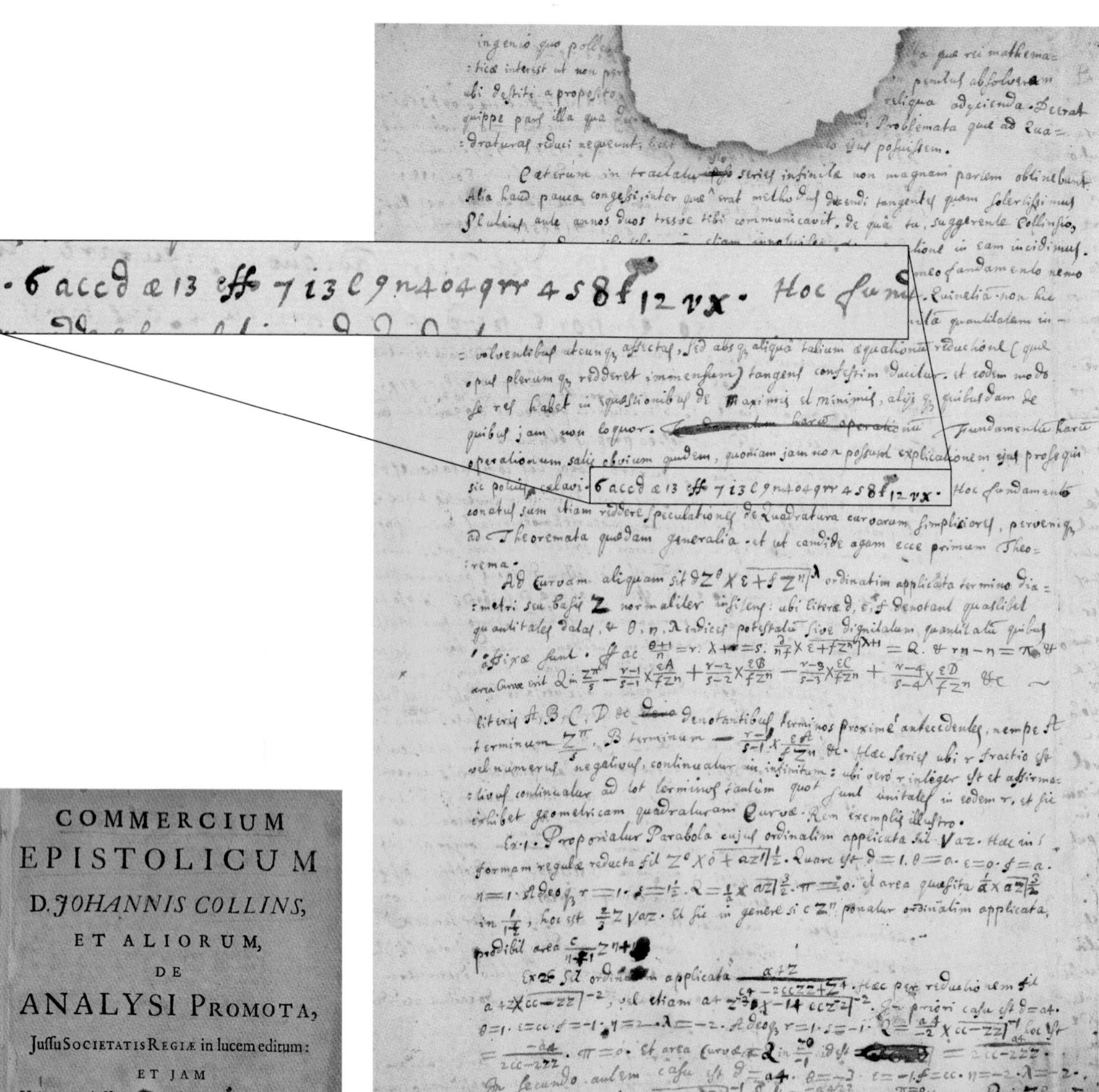

6accdæ13eff7i3l9n4o4qrr4s8t12vx. Hoc fund

COMMERCIUM
EPISTOLICUM
D. JOHANNIS COLLINS,
ET ALIORUM,
DE
ANALYSI PROMOTA,
Jussu Societatis Regiæ in lucem editum:
ET JAM
Unà cum ejusdem Recensione præmissa, & Judicio primarii, ut ferebatur, Mathematici subjuncto, iterum impressum.

LONDINI:
Ex Officinâ & impensis J. Tonson, & J. Watts, prostant venales apud Jacobum Mack-Euen, Bibliopolam Edinburgensem. M DCC XXII.

The *Commercium epistolicum* (1712) – the report of a committee adjudicating the priority dispute – reproduced a key piece of evidence establishing Newton's priority: an anagram that Newton included in a letter to Leibniz (above) in October 1676 and in which he concealed his formulation of the fundamental theorem of the calculus. – Letter: Cambridge University Library; *Commercium epistolicum*: NYPL–SIBL

tion made Newton unwilling to consider anything short of bringing Leibniz to his knees. As the German was unwilling to plead his case, Newton was perfectly happy to do it for him. From his Chair, the Royal Society President saw to it that an "impartial" committee was appointed on March 6, 1712, to look into the matter and return a report. The committee members were mostly staunch Newtonians, such as Edmond Halley, William Jones, John Machin, Abraham de Moivre, and Brooke Taylor. For good measure, the Prussian envoy to London was added a week before the report was submitted.

For all we know, members of the committee met and sifted through documents provided by Newton or by Sloane. (Leibniz's version did not matter.) They needn't have bothered; Newton took it upon himself to carry out the research, as well as to author the report, which was presented to the Royal Society on April 24: "We reckon Mr Newton the first Inventor," was the unanimous verdict; "and are of opinion that Mr Keill in asserting the same has been noways injurious to Mr Leibniz." The report was soon printed under the title *Commercium Epistolicum D. Johannis Collins et aliorum de analysi promota* ("The Correspondence of John Collins and Others About the Development of Analysis"), and sent privately to influential savants at home and abroad. To ensure that the message would not get lost, Newton also reviewed the book (anonymously, of course) for the *Philosophical Transactions.*

The impact of the *Commercium Epistolicum* on the European scientific community was considerable. Overnight, Leibniz's nearly enshrined stature as the sole discoverer of the calculus was severely fractured. The force of the considerable documentation marshaled by Newton was almost stupefying. Leibniz's closest friends, Johann Bernoulli and Christian Wolff, urged Leibniz to publish a historical narrative of his own, lest "most people may deduce from silence that the English case is a good one." But there was little that Leibniz could do. Newton *was* able to present a convincing case for his prior discovery of the calculus – even if he proved less convincing on the matter of plagiarism. In contrast, Leibniz's private papers were in no shape to be made public as proof of the progress he had made in his studies. Thus, after an abortive start, he abandoned the idea. He did publish, however, a short rebuttal of the *Commercium* in June 1713, the *Charta volans* ("Fly-sheet"). In it, he went beyond a mere statement of his case; he openly charged Newton with willfully lifting the calculus from him.[37]

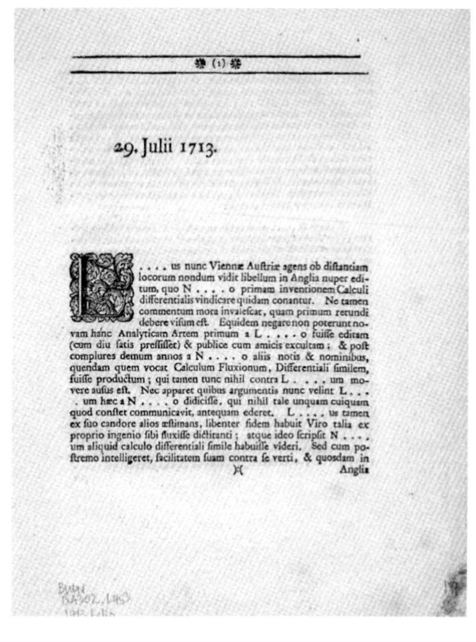

(1)

29. Julii 1713.

L. . . . us nunc Viennæ Auſtriæ agens ob diſtantiam locorum nondum vidit libellum in Anglia nuper editum, quo N . . . . o primam inventionem Calculi differentialis vindicare quidam conantur. Ne tamen commentum mora invaleſcat, quam primum retundi debere viſum eſt. Equidem negare non poterunt novam hanc Analyticam Artem primum a L . . . . o fuiſſe editam (cum diu ſatis preſſiſſet) & publice cum amicis excultam; & poſt complures demum annos a N . . . . o aliis notis & nominibus, quendam quem vocat Calculum Fluxionum, Differentiali ſimilem, fuiſſe productum; qui tamen tunc nihil contra L . . . . um movere auſus eſt. Nec apparet quibus argumentis nunc velint L . . . . um hæc a N . . . . o didiciſſe, qui nihil tale unquam cuiquam quod conſtet communicavit, antequam ederet. L . . . . us tamen ex ſuo candore alios æſtimans, libenter fidem habuit Viro talia ex proprio ingenio ſibi fluxiſſe dictitanti; atque ideo ſcripſit N . . . . um aliquid calculo differentiali ſimile habuiſſe videri. Sed cum poſtremo intelligeret, facilitatem ſuam contra ſe verti, & quoſdam in

)(Anglia

In June 1713, Leibniz published a short rebuttal of the *Commercium*, the *Charta volans* ("Fly-sheet"). In it, he openly charged Newton with willfully lifting the calculus from him. – Burndy Library

The dispute continued, unabated, for another three years, until Leibniz's death in 1716. By then, the intensity of the enmity between the two warring parties was immeasurable. An anecdote perhaps captures it best. According to William Whiston, Newton was once overheard "pleasantly" telling Samuel Clarke that "he had broke Leibnitz's Heart with his Reply to him."[38] A cruel statement indeed;

The Swiss mathematician Johann Bernoulli (1667–1748) was a friend and staunch champion of Leibniz in the priority dispute. – Smithsonian Institution Libraries, Washington, D.C.

but Newton became a man possessed as the debate metamorphosed into a theological and metaphysical debate – and on these grounds Newton was unforgiving.

Its ramifications went well beyond any personal toll it exacted on the immediate participants, auguring lasting consequences for the European scientific community in general, and for the diffusion of Newtonian ideas in particular. First, it pitted against each other not just ideas and individuals, but nations. Even before the dispute, French and German savants had complained of what they perceived to be a British proclivity to magnify the accomplishments of their compatriots and to overlook the contributions of foreigners. Now, the rancor spawned by the priority dispute hardened the general xenophobic outlook. Henceforth, acceptance or rejection of a position was as likely to be made along nationalistic lines as on the merits of the case. Already, in his first major attack on Leibniz, John Keill had proclaimed that he was entitled to his critique on the grounds that the Germans had so magnified Leibniz that it was only "proper for Britons to restore to Newton what was snatched from him." A fellow-Newtonian, William Jones, concurred. "The Germans and French have in a violent manner attacked the Philosophy of Sir Is. Newton," he wrote in October 1711, "and seem resolved to stand by Cartes; Mr Keil, as a person concerned, has undertaken to answere and defend some things, as Dr Friend, and Dr Mead, does (in their way) the rest." Newton himself found it convenient to adopt such a simplistic view of the dispute. When, in May 1712, he considered responding to Leibniz's accusation that

Newton resorted to supernatural, rather than mechanical, explanations for gravity, Newton represented the attack as "reflecting upon the English." As far as he was concerned, there was an English school committed to (his) theory of universal gravitation; such a commitment, with its attendant refusal to "explain gravity by a mechanical hypothesis," brought upon them charges of embracing "a supernatural thing, a miracle, a fiction invented to support an ill grounded opinion."[39]

Closely connected to the hardening of positions along nationalistic lines was the loss of good will among the Continental mathematicians, who, almost to a person, were disciples of Leibniz or, more accurately, of his staunch lieutenant, Johann Bernoulli. The latter, during the 1690s, was only slightly less admiring of Newton and the *Principia* than Leibniz himself. With the unfolding of the priority dispute, Bernoulli emerged to stand for Leibniz as Keill stood for Newton, and his unrelenting animus (and longevity) boded ill for the European reception of Newtonian physics. The shape of things to come was clear to all as soon as the dispute erupted. Shortly after Leibniz filed his complaint against Keill with the Royal Society, Bernoulli began speculating about which of the errors Leibniz had found in the *Principia* would be corrected by its author in the second edition. Bernoulli himself had discovered one careless error in Proposition 10 of Book II – which he communicated to Leibniz in August 1710 – and, assuming that it would not be detected by the British, Bernoulli intended to use it to expose Newton the instant the second edition appeared. As it happened, Bernoulli's own nephew played spoiler. Nikolaus Bernoulli visited England in the fall of 1712 and informed Newton of the error. Though the relevant part of the book was already printed, Newton managed to recast the entire section and escape lasting embarrassment. However, Bernoulli's ploy was now laid bare: what was at stake was Newton's very mathematical competency. As has been observed, hereafter that competency "was assailed and every effort was made to convict [Newton] of error and ignorance on the ground that so feeble a mathematician could not conceivably have devised the calculus."[40]

3

In "The Ancient of Days," the frontispiece to his *Europe, a Prophecy* (1794), William Blake depicts the powerful god Urizen (a pun on "your reason"), a heroic figure fashioned on Newton who, kneeling on his right knee, "measure[s] out the immense," and with his vast compasses creates a "world better suited to obey His Will." For Blake, however, such a rationalist act of creation is lifeless, for its abstraction and materialism severely limit divinity and imagination. - NYPL-Berg Collection

TRIAL BY FIRE

Even as the priority dispute gained momentum, Newton in 1709 allowed himself to be drawn into an immensely challenging (and rewarding) collaboration with the Cambridge mathematician Roger Cotes, which led to the publication of the second edition of the *Principia* in 1713. No doubt, Newton's perception of Leibnizian recalcitrance rankled, feeding off the mental energy he had expended over the years to reconstruct the creative process that had given birth to his masterpiece in 1687. So, too, did his indignation with his old nemesis, Descartes. The fusillade opened by John Keill, then – if not engineered by Newton – was animated by Newton's determination to rout his opponents, both past and present. Within this context, the revised *Principia*, under Cotes's editorship, served as a declaration of war on Cartesianism. The French Cartesians, who had hitherto paid relatively little attention to the central physical claims of the *Principia* – and whose attention during the past two decades had been focused on the calculus rather than on gravity – responded vigorously; by 1720, a distinct split had emerged between Cartesians and Newtonians in France, mirroring the split between Leibnizians and Newtonians in central Europe.

The reverberations of these giant rifts in the scientific community were soon felt throughout all of Europe.

Newton refused to pen the preface to the revised *Principia*, or even to read the preface finally authored by Cotes, which began with a scornful dismissal of the Aristotelians, whom he saw as preoccupied with "philosophical jargon" and thus incapable of contributing to the study of nature. The mechanical philosophers who followed on their heels, Cotes continued, began on firmer ground but ended up equally misguided, guilty of "imagining that the unknown shapes and sizes of the particles are whatever they please, and of assuming their uncertain positions and motions, and even further feigning certain occult fluids that permeate the pores of bodies very freely, since they are endowed with an omnipotent subtlety and are acted on by occult motions." "Occult qualities" were invoked by Cotes in order to counter charges the Leibnizian camp had recently made vis-à-vis the occult nature of universal gravitation. Leibniz had already voiced the charge privately on several occasions in the early years of the eighteenth century, and went public with the charge in the *Theodicy* of 1710 and in the 1711 open letter to the anti-Newtonian naturalist Nicolaus Hartsoeker that included a gratuitous attack on Newton: if one claims that attraction is something that requires no mechanism, and exists simply by the decree of God, "who produces this effect without employing any intelligible means," Leibniz thundered, "it is a senseless occult quality, which is so very occult, that it is impossible it can ever be cleared up, even though an angel, nay, God himself, would attempt to explain it." Ultimately, it was the anti-Newtonian camp that garnered a monopoly on this term of abuse, which soon emerged as a key weapon in its arsenal.[1]

Roger Cotes (1682–1716), first Plumian Professor of astronomy and experimental philosophy at Cambridge, edited the second edition of the *Principia*. After Cotes's untimely death at age thirty-four, Newton commented that, had he lived, "we might have known something." – Trinity College, Cambridge

Although Newton did not object to Cotes's endeavor to charge the Cartesians with propounding occult qualities, he prevented Cotes from openly attacking Leibniz's *Tentamen*. As a result, Cotes remained focused on the Cartesians. When the mechanical philosophers admit such occult motions, he continued,

> they are drifting off into dreams, ignoring the true constitution of things, which is obviously to be sought in vain from false conjectures, when it can scarcely be found out even by the most certain observations. Those who take the foundation of their speculations from hypotheses, even if they

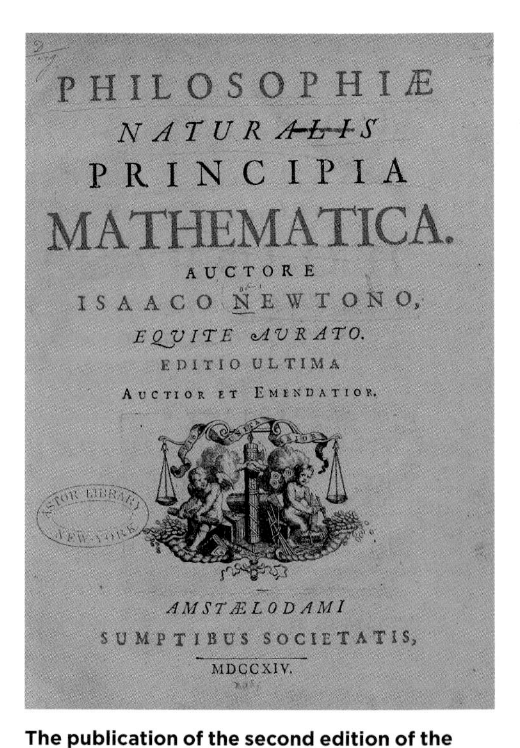

PHILOSOPHIÆ
NATURALIS
PRINCIPIA
MATHEMATICA.
AUCTORE
ISAACO NEWTONO,
EQUITE AURATO.
EDITIO ULTIMA
AUCTIOR ET EMENDATIOR.

AMSTÆLODAMI
SUMPTIBUS SOCIETATIS,
MDCCXIV.

The publication of the second edition of the *Principia* in 1713 spurred the diffusion of Newton's masterpiece on the Continent, hand in hand with the dissemination of the Leibnizian calculus. Shown here is the pirated Amsterdam edition of 1714. – NYPL–SIBL

then proceed most rigorously according to mechanical laws, are merely putting together a romance, elegant perhaps and charming, but nevertheless a romance.

Cotes proceeded to describe the true path to natural philosophy: the method of analysis and synthesis by which Newton "was the first and only one who was able to demonstrate [universal gravity] from phenomena and to make it a solid foundation for his brilliant theories."[2]

The assault on vortices and other Cartesian tenets was mirrored in Newton's own offensive in the body of the work. In the revised edition of the *Principia,* Newton replaced the term "hypothesis" with three "Rules for the Study of Natural Philosophy" calculated to evoke (and demean) Descartes' recently published manuscript "Rules for the Direction of the Mind." More significant in this context was the new conclusion added to the *Principia* – the "General Scholium" – wherein Newton addressed topics that he had toned down or suppressed altogether in 1687. He now categorically trounced vortices; he openly postulated a design argument drawn from phenomena of nature rather than from a-priori reasoning; and he asserted – defiantly – that though he was ignorant of the *cause* of gravity, yet it was a fact and, further to the point, it did not act "as mechanical causes are wont to do." It was here that Newton propounded famously on hypotheses as well as attempted to hurl back at his opponents the loaded metaphor "occult qualities":

I have not as yet been able to deduce from phenomena the reason for these properties of gravity, and I do not feign hypotheses. For whatever is not deduced from the phenomena must be called hypothesis; and hypotheses, whether metaphysical or physical, or based on occult qualities, or mechanical, have no place in experimental philosophy.[3]

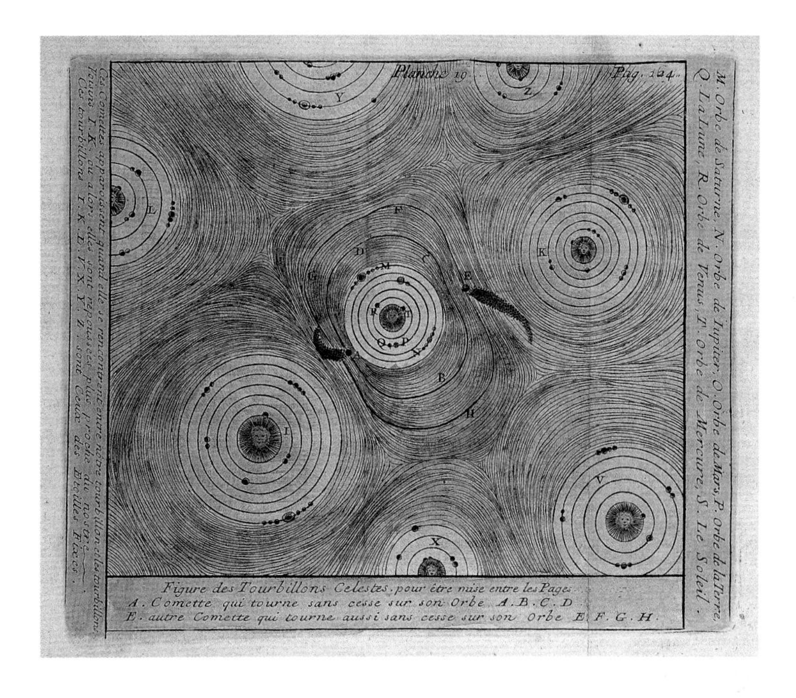

Offering an attractive representation of the Cartesian universe, Nicolas Bion's "The Shape of Celestial Vortices" – from his *L'usage des globes céleste et terrestre* (Paris, 1752) – attempts to ascribe orbital motion to comets (drawn in red). – Courtesy of Adler Planetarium & Astronomy Museum, Chicago, Illinois

The Reception in France

The revised *Principia* was published in June 1713; within a year, a pirated edition appeared in Amsterdam, suggesting a demand for copies on the Continent. Newton himself sent presentation copies to key members of the French Académie des Sciences, including the abbé Bignon, who frequently served as its president. Responding to Bignon's acknowledgment of the gift, Newton could not resist taking a jab at the likely French response to a book that was ostensibly ill-suited to their predilection for conjectures. Only a few would be able to read the book, he predicted, partly because of its recondite nature and partly because the matters treated therein "have been made rather obscure by excessive brevity." Nonetheless, Newton gleefully prognosticated that what would prove most objectionable about the *Principia* was that such matters "run counter to the philosophical hypotheses commonly received, and [also] are rendered the more unwelcome by the lack of hypothetical explanations." Newton's overall prediction proved accurate.

Pierre Varignon (1654–1722) was professor of mathematics at the Collège Mazarin (and later at the Collège de France), and a member of the Académie des Sciences. He was instrumental in the diffusion in France of both Newtonian mechanics and the Leibnizian calculus. – Smithsonian Institution Libraries, Washington, D.C.

Though the debates were only beginning to unfold, Newton's standing as a towering intellect was never in question. There exists no better testament to his stature in France already by the late 1690s than the attempt to recruit him as a member of the Académie. Within months of the signing of the Treaty of Ryswick in September 1697 – which brought to a close the first phase of the Anglo/Dutch-French war – Jacques Cassini traveled to England, partly, it appears, to offer Newton an appointment as a salaried academician. Newton declined the Académie's offer, but in 1699 he accepted his nomination as foreign associate there. Such early esteem of Newton was articulated in 1713 by Bernard de Fontenelle, the Académie's Secretary, when he thanked Newton for a copy of the revised edition. For several years, he wrote, Newton's "excellent" *Principia* had been admired in Europe, but nowhere more so than in France, where one truly appreciates the merits of foreigners.[4]

As noted above in chapter 2, the first edition of the *Principia* was greeted with genuine interest among the cognoscenti on the Continent – Huygens, Leibniz, and Johann Bernoulli, in particular. Their estimation of the text could be characterized as one of reverential disagreement, combined with a determination to probe the issues under contention and engage the author in the process. The initial French reception followed similar lines. The first public debate over the *Principia* appears to have taken place on April 27, 1690, during the visit of the exiled James II to the Paris Observatory. Exhibiting familiarity with some aspects of a book that was, after all, dedicated to him, the deposed monarch lectured his hosts on Newton's conclusions regarding the effects of gravity on the shape of the earth. The presentation generated a lively debate, with the academicians respectfully rejecting Newton's theory. The topic remained a bone of contention for decades, with the Observatory astronomers proving the most vociferous opponents of Newton's notions regarding the oblate shape of the globe.[5]

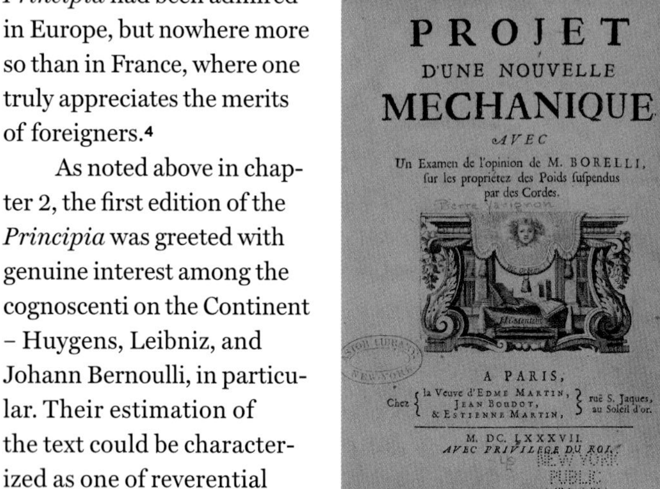
PROJET
D'UNE NOUVELLE
MECHANIQUE
AVEC
Un Examen de l'opinion de M. BORELLI, sur les proprietez des Poids suspendus par des Cordes.
A PARIS,
Chez la Veuve d'EDME MARTIN, JEAN BOUDOT, & ESTIENNE MARTIN, ruë S. Jaques, au Soleil d'or.
M. DC. LXXXVII.
AVEC PRIVILEGE DU ROY.

In his first publication, *Projet d'une nouvelle mechanique* (Paris, 1687), Varignon advertised his own research program, which was remarkably similar to the one articulated in Newton's *Principia*. In the years that followed, Varignon contributed greatly to the emergence of rational mechanics. – NYPL–SIBL

One recently inducted academician who must have attended the royal visit was Pierre Varignon. Coming from a poor family of masons, Varignon was tonsured at the age of twenty-two, and his proclaimed vocation made it possible for him to attend the university at Caen. He became a priest, but in 1686 he accepted the invitation of his friend the abbé de St. Pierre to move with him to Paris. The following year, contemporaneous with the publication of the *Principia*, Varignon brought out a short treatise, *Projet d'une nouvelle*

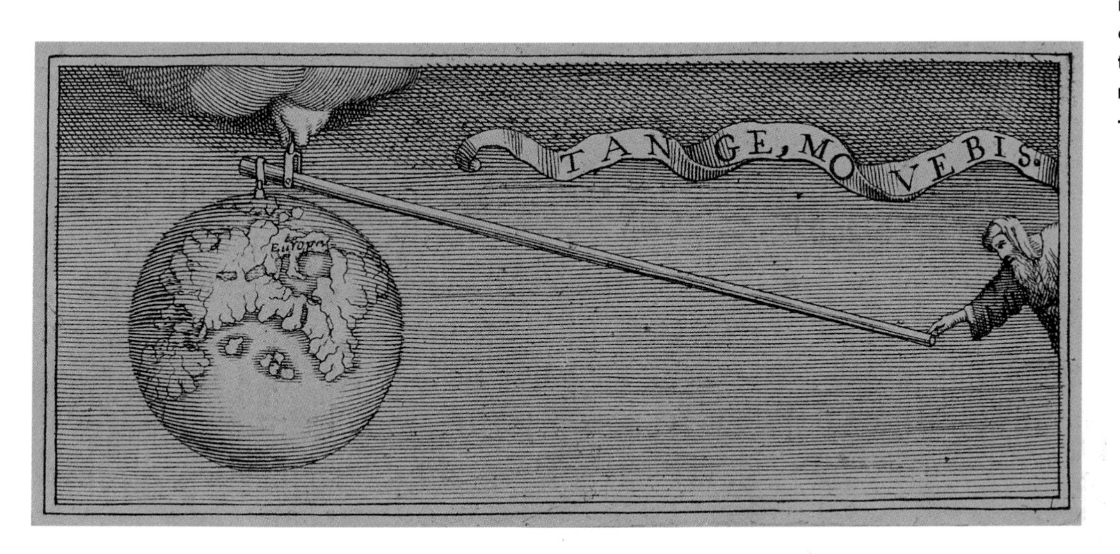

This vignette from Varignon's *Projet* offers a striking visual representation of the famous statement by Archimedes: "Give me a lever long enough and a place to stand, and I will move the world." - NYPL-SIBL

mechanique ("Project for a New Mechanics"), in which he articulated the contours of a research program remarkably similar to that articulated in Newton's masterpiece. Coming on the coattails of the *Principia*, Varignon's *Projet* undoubtedly garnered added attention, resulting in its author's election in 1688 as Professor of Mathematics at the Collège Mazarin and his membership in the Académie des Sciences.

Writing in 1713 to thank Newton for a presentation copy of the revised *Principia*, Varignon waxed eloquent on how, a quarter of a century earlier, he had been roused to "burning eagerness" by his encounter with the first edition of Newton's "exquisite" treatise, which he read "with exceedingly great pleasure, wondering always at the supreme strength and sharpness of your genius, with which you unlock the door of nature, penetrate into her inmost recesses, and most skilfully demonstrate by reasoning derived from a sublime geometry, laws which no one who was not most acute could perceive, which you have discovered there." Varignon admitted that perusing Newton's magnum opus around 1688 had "provoked many new ideas in my mind; by adding these to the margins of your extraordinary book, I have stained all of them completely with notes of this sort, for what they are worth."[6] The effusive rhetoric notwithstanding, the French abbé comes off as quite candid in his account of discovering in Newton a kindred spirit and a fellow traveler in the exciting new science of mechanics. Varignon's immersion in the *Principia*, as well as in the Leibnizian calculus, would continue unabated until his death; his writings and teaching would prove as crucial for the development of analytical mechanics as they would for the rising reputation of Newton in France.

Newton's repudiation of analysis, and his consequent "failure" to use "fluxions" in the *Principia*, have led some scholars to argue that Newton should not be credited with the founding of rational mechanics; it was owing

Sébastien Le Clerc's *Nouveau systême du monde* (Paris, 1706) was one of many French attempts in the early eighteenth century to explicate the system of the world roughly along Cartesian lines. His exposition of a system of vortices was advertised as conforming to Holy Scripture. – NYPL–SIBL

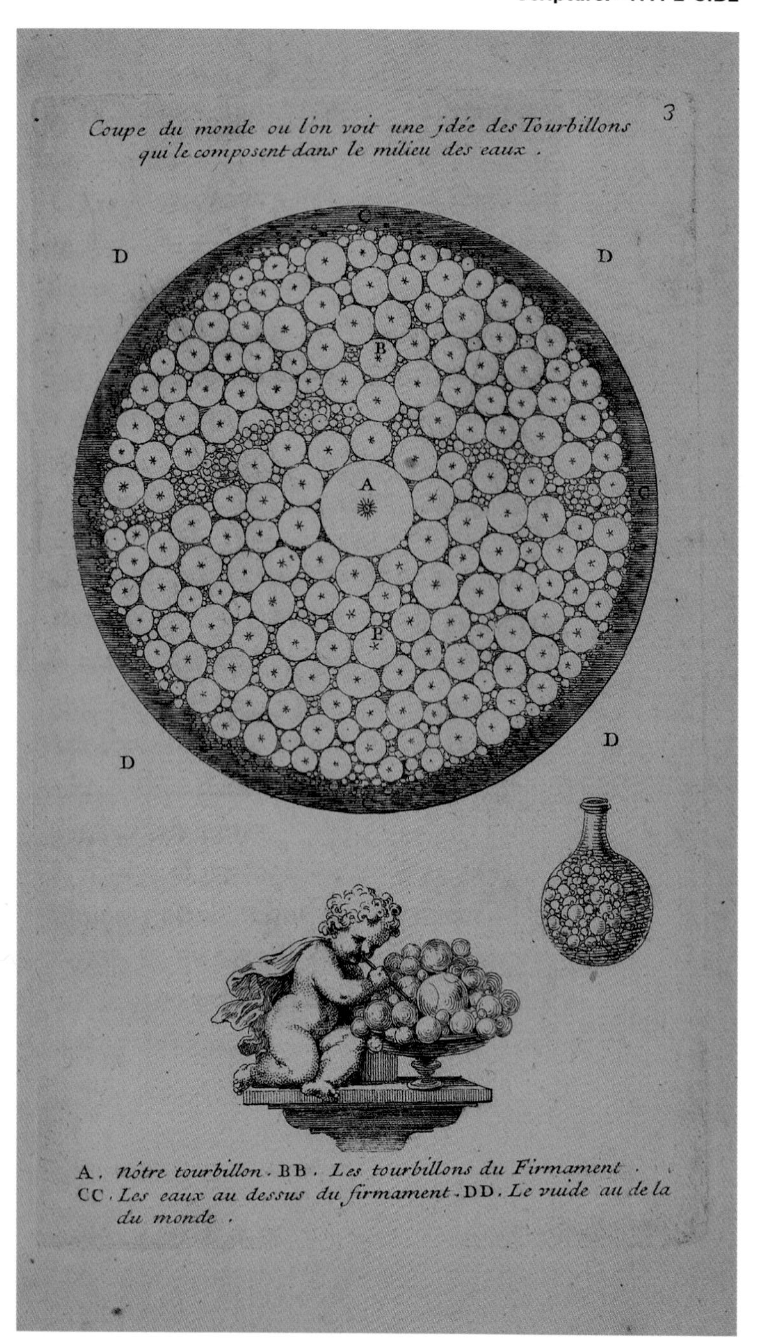

primarily to the work of Frenchmen, from Varignon to the mathematician Joseph-Louis Lagrange, that such a discipline came into being. Without denying the merit of this retrospective appreciation, one must keep in mind that Newton declared openly that the creation of such a science was precisely what he intended: "*rational mechanics* will be the science, expressed in exact propositions and demonstrations, of the motions that result from any forces whatever and of the forces that are required for any motions whatever."[7] Yes, his successors rendered Newton's work into differential equations, clearing up errors and obscurities along the way; but all along, notwithstanding the (justified) pride they took in their contributions, they remained sensible that they were following Newton's lead. Thus, when Varignon lectured on central forces before the French Académie in the mid-1690s, his debt to the *Principia* was obvious to him as well as to his discerning auditors. The same insight applies to many Continental mathematicians trained on Leibnizian principles: though they may have carried to an excess the abstract rendering of analytical mechanics, or may have handled central forces as mathematical relations divorced from reality, their license to do so derived from Newton – no matter how the great man might have shuddered at their wanton divorce of mathematics from physics.

Be this as it may, Varignon faced an uphill battle in his efforts on behalf of analytical mechanics. By the mid-1690s, a formidable opposition to the new calculus had emerged within and outside the Académie des Sciences, to the extent that as early as 1697 Varignon portrayed himself as "a true martyr" to the cause. His only ally at the time was the Marquis de l'Hôpital, who in 1696 published an influential textbook, the *Analyse des infiniment petits*. This dispute with those adamant that the new mathematics lacked both rigor and a proper foundation continued until 1706, when the "peace of the infinitely small" was forced on the opponents of the calculus. Significantly, at one point in the debate Varignon resorted to invoking the *Principia* to demonstrate the rigorous foundations of the calculus![8] When the dust finally settled, not only did the calculus rest on more secure philosophical foundations, and its rigor been demonstrated, but the applications of the new differential equations had multiplied beyond what even Leibniz might have predicted in 1684 when he published his groundbreaking paper.

No sooner was peace restored to the French Académie than, with the publication in 1707 of Philippe Villemot's *Nouveau système, ou, Nouvelle explication du mouvement des planètes*, the focus of attention shifted back to celestial mechanics. Villemot's radial impulse theory, which sought to reconcile vortices with Kepler's laws – the "Achilles heel" of the Cartesian system exposed in the *Principia* – was developed in blissful ignorance of the recent developments in the field; even the debates over the calculus that had engulfed the French community in the previous decade seemed to have eluded Villemot. Certainly, he did not mention the *Principia* in the text (his preface excused the omission of the "clever Englishman" on the grounds that the *Principia* was hard to get). Only after he had completed his manuscript, Villemot contended, did he manage to get hold of a copy, but still he found no reason for revision.

Fontenelle, the perpetual Secretary of the French Académie des Sciences, enthusiastically reviewed Villemot's creative effort to rescue Descartes, even though the author – as Leibniz would later sneer – sought to explain celestial motion without availing himself of a single demonstration. Nor was Bernoulli or Varignon any more pleased than Leibniz was with Villemot or, for that matter, with Fontenelle. What informed the latter's reaction, however, was a penchant to preserve as much as possible of Cartesian physics. Indeed, in Fontenelle's annual summaries of the work carried out by members of the Académie, he routinely subverted the import of Varignon's researches on central forces to suit this overarching goal. Whereas the latter deliberately steered clear of any discussion of vortices, to the point of neutrality in regard to universal gravitation (and, more troublesome still, not only relied heavily on the *Principia* but treated centripetal as well as centrifugal forces), Fontenelle systematically made Varignon's results appear to dovetail with the broad outlines of Cartesian theories.[9]

Such attempts by Fontenelle to force Varignon into a Cartesian mold were complicated not only by the evolving nature of Varignon's work, but by the proliferation of efforts to establish vortices on more secure grounds. The Academician Joseph Saurin, who, according to Fontenelle, "appeared always strongly convinced that the true philosophers should use all their efforts to conserve the vortices of Descartes," posited a more mathematically rigorous – and experimentally cogent – approach than Villemot. His endeavor to make Descartes' vortices conform

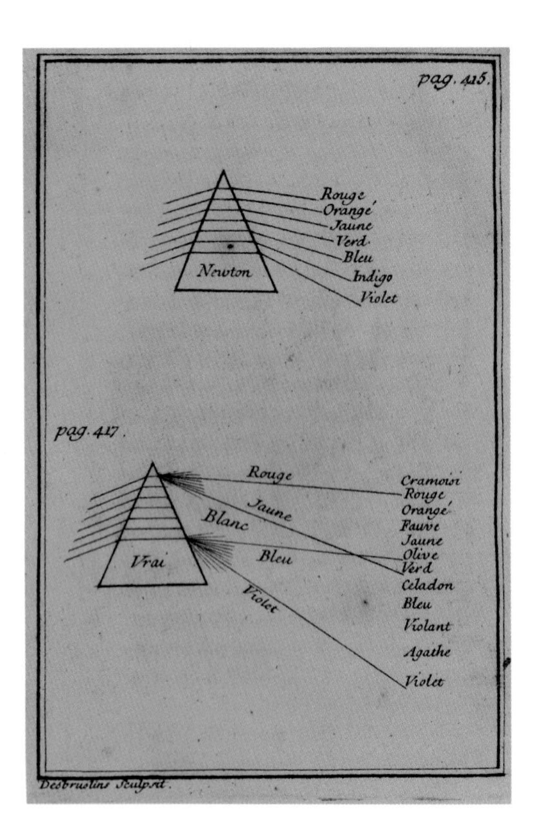

This diagram from Louis-Bertrand Castel's *L'optique des couleurs* (Paris, 1740), in which he described his invention of a color harpsichord, contrasts Newton's spectrum with Castel's "true" system. - NYPL-SIBL

to Kepler's laws led him to propound the existence of an extremely rarefied subtle fluid (aether), which posed no perceptible resistance to planetary motion. Nothing in his solution, Saurin declared in his 1709 paper, diminished the exactitude of the *Principia*. By his own admission, Newton was untroubled by the fact that "his suppositions are not reasonable," and proceeded "to consider *pesanteur* [gravity] as an inherent quality of bodies." As far as Saurin was concerned, such a casual approach "only revived ideas like attraction and occult qualities that have rightly been decried." "Let us always use clear mechanical principles in our reasoning," Saurin concluded. "If we abandon them, all the light we can hope to achieve is extinguished, and we will once again be returned to the ancient darkness of Peripatism from which heaven wants to save us."[10]

Saurin's efforts to reconcile vortices with Kepler's laws, had they been followed, might have helped to bring the Cartesian and Newtonian positions closer a generation earlier. But in 1712 Saurin's aether theory was rejected by the venerable Oratorian philosopher Nicolas Malebranche who, in his sixth edition of the *Recherche de la vérité* (1712), reverted back to Descartes' dense medium and sought to improve on Villemot's theory – with the same disregard of work carried out during the previous three decades. For the short term at least, Malebranche's interpretation carried the day in France, and this neo-Cartesian orthodoxy consolidated a scientific opinion – just as copies of the contentious new edition of the *Principia* were making their way across the channel.

Ironically, Malebranche's last-ditch efforts to salvage Cartesian vortices followed his partial conversion to Newton's optics and mathematics. In 1699, upon his election to membership in the Académie des Sciences, Malebranche propounded an optical theory that conceived of light as vibratory motion of subtle matter. Different frequencies of these pulses generated three primary colors (red, yellow, blue), the mixture of which produced the other hues. When the English essayist Joseph Addison visited Malebranche in 1700, the Oratorian entertained him by reading this "very pretty Hypothesis of Colours," different from either Descartes' or Newton's. Malebranche, however, appears to have added the caveat that "they may all three be true."[11] Upon the publication of Newton's *Opticks* in 1704, Malebranche began reevaluating his positions. The mathematical treatises Newton appended to the English edition led him to conclude that Newton "pushes the integral calculus farther than everything that has been printed up to now." This newly acquired high regard did not prevent Malebranche from maintaining in the same year that Newton, though an "excellent mathematician," was no "physicist." He judged the *Opticks* to be "very curious and very useful to those who have the right principles in physics," further commenting on the "principal experiments that M. Newton, this learned mathematician, so famous in England and everywhere, has made with an exactitude such that I cannot doubt their truth." Significantly, Malebranche made public his conversion to Newton's theory of colors –while retaining his commitment to wave theory – in the sixth edition of the *Recherche de la vérité* (1712), wherein he also propounded his doctrine of small vortices.[12]

Whether Malebranche verified Newton's experiments himself is unclear, but his enthusiasm certainly infected others. Thus, for example, his 1714 correspondence with the provincial natural philosopher Jean-Jacques Dortus de Mairan stimulated the latter to undertake a series of experiments during 1716–17 that confirmed Newton's results and made de Mairan a convert. Malebranche may also have had a hand in arranging for the chemist Etienne-François Geoffroy to present the French Académie with a précis of Newton's *Opticks*.[13]

It appears, however, that at least some members of the French Académie were convinced that Newton "despised" the distinction conferred upon him in 1699 when he was elected foreign associate, and this may have fortified Newton's resolve to counter the proselytizing efforts of the Leibnizians by mounting a campaign of his own. The campaign opened, as noted above, in 1713 with the dispatch of copies of the second edition of the *Principia* to key members of the Académie: President Bignon, Secretary Fontenelle, and Director of Research Varignon. In early 1715, when a delegation from the French Académie visited England to observe the total solar eclipse, Newton took advantage of the opportunity to admit its members – astronomer Jacques-

The desire to observe the total solar eclipse of April 22, 1715 – predicted by Edmond Halley and seen here in a drawing by Roger Cotes – drew to England a delegation from the French Académie des Sciences. Newton took advantage of the occasion to induct the delegates as Fellows of the Royal Society and to entertain them with the replication of his prismatic experiments.
– Trinity College, Cambridge, MS R. 16.38, fols. 293–4

NEWTON'S COLORS AND PAINTERS' COLORS

1

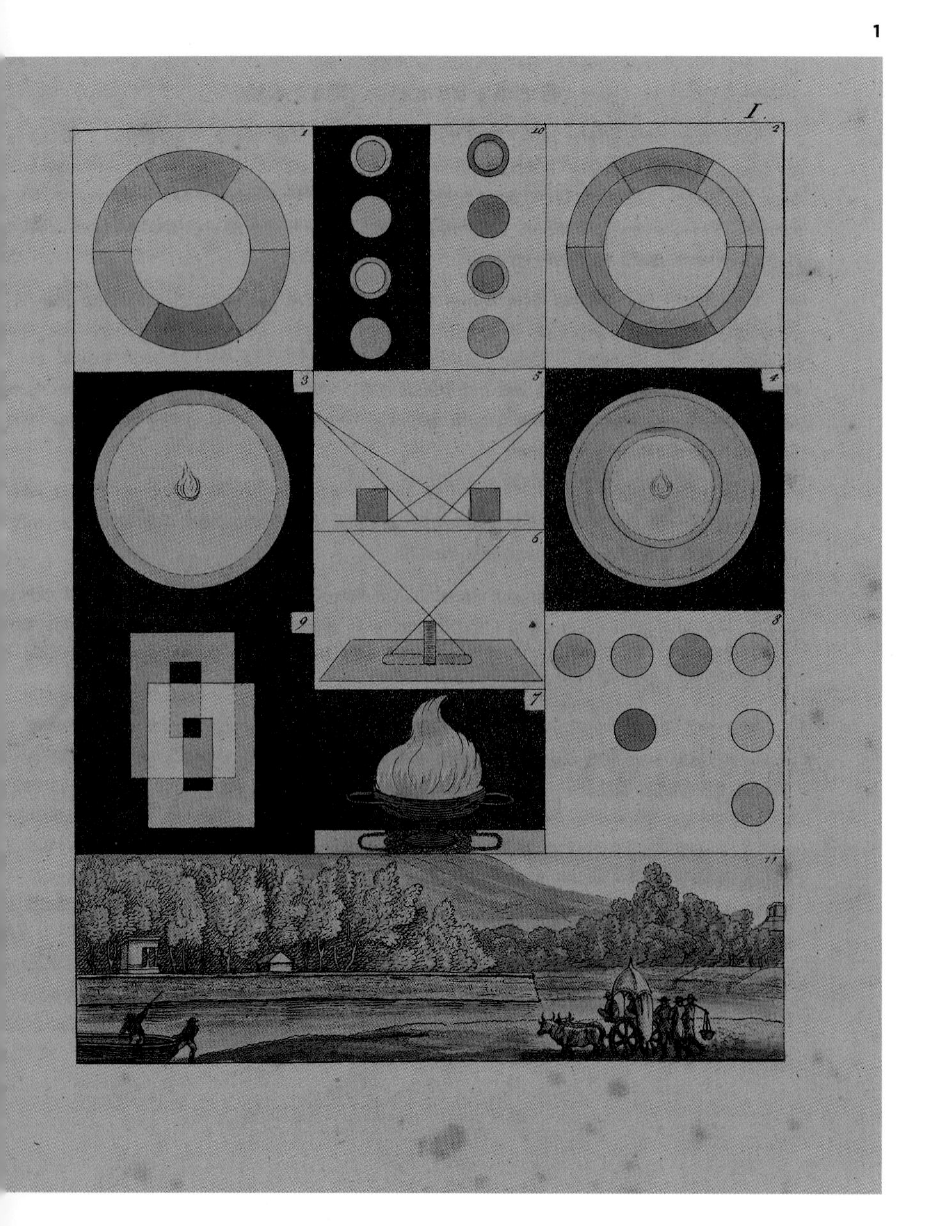

Newton's use of the common language of painters to explain his theory of colors proved to be a continuing source of misunderstanding. Newton had distinguished between colors produced by mixing lights and those produced by mixing paints or colored powders. In the *Opticks*, he asserted that only seven primary colors were required to generate all the others by the mixing of lights, and he drew a color circle to indicate the order and proportion of the different hues involved. Painters, however, failed to grasp Newton's fundamental distinction between mixed lights and mixed paints. This exacerbated the conflict over which were the true primary colors and whether Newton's fundamental claim that white was a mixture of all colors could be correct (since paints, unlike lights, cannot be mixed to produce white). Shown on these two pages are plates representing Goethe's vehement refutation of Newton's color theory [1]; three-color systems purported to have been inspired by Newton [3a–b; 5]; Schiffermüller's "Newtonian" twelve-color system [4]; and Runge's mystical color theory [2], partly conceived in response to Goethe's theories.

2

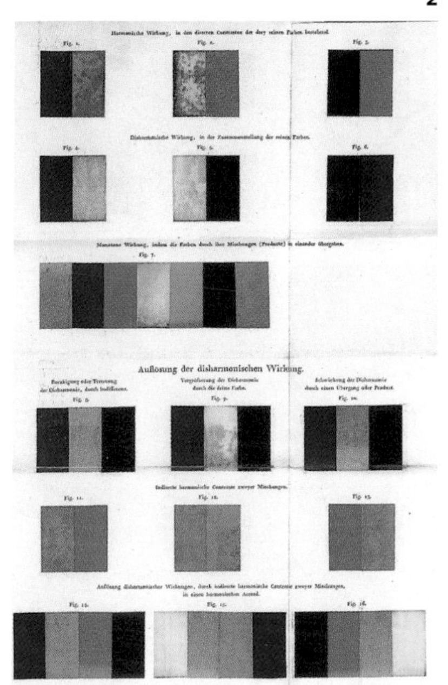

3a

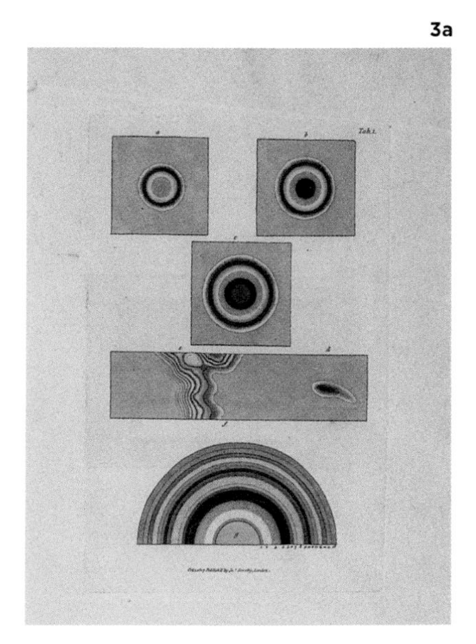

3b

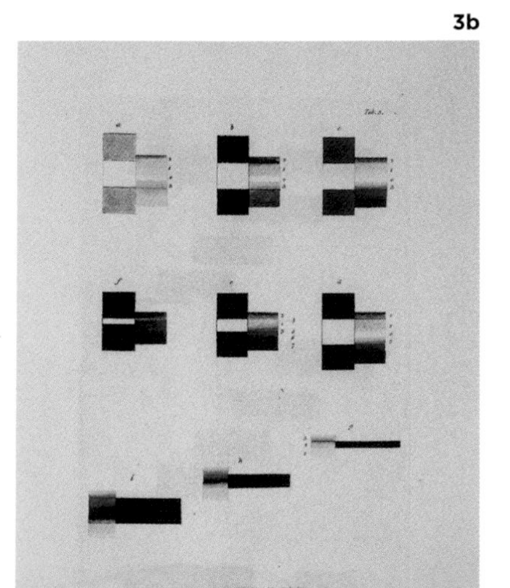

4

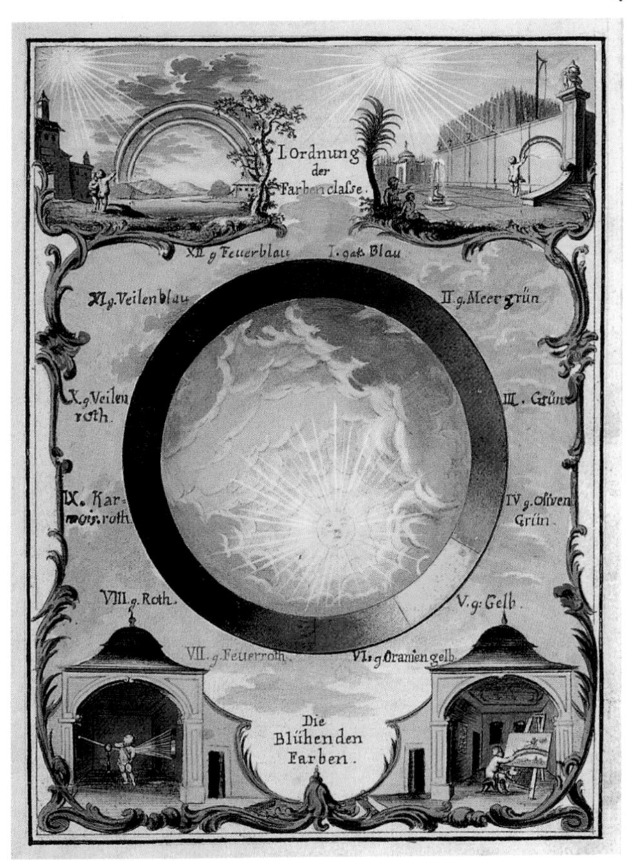

5

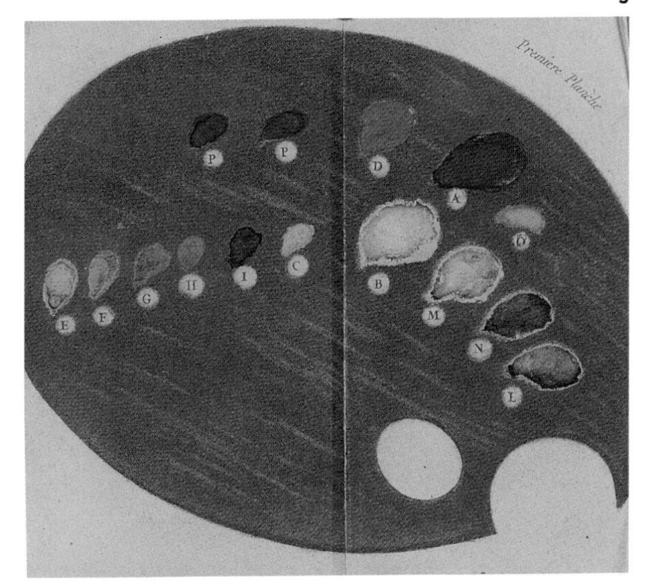

1: Johann Wolfgang von Goethe, ***Zur Farbenlehre*** **(Tübingen, 1810) – NYPL–SIBL**

2: Philipp Otto Runge, ***Farben-Kugel*** **(Hamburg, 1810) – Harvard College Library**

3a–b: James Sowerby, ***A New Elucidation of Colours, Original, Prismatic, and Material*** **(London, 1809) – Burndy Library**

4: Ignaz Schiffermüller, ***Versuch eines Farbensystems*** **(Vienna, 1771) – The Metropolitan Museum of Art, Thomas J. Watson Library**

5: Antoine Gautier de Montdorge, ***L'art d'imprimer les tableaux*** **(Paris, 1756) – NYPL–Print Collection**

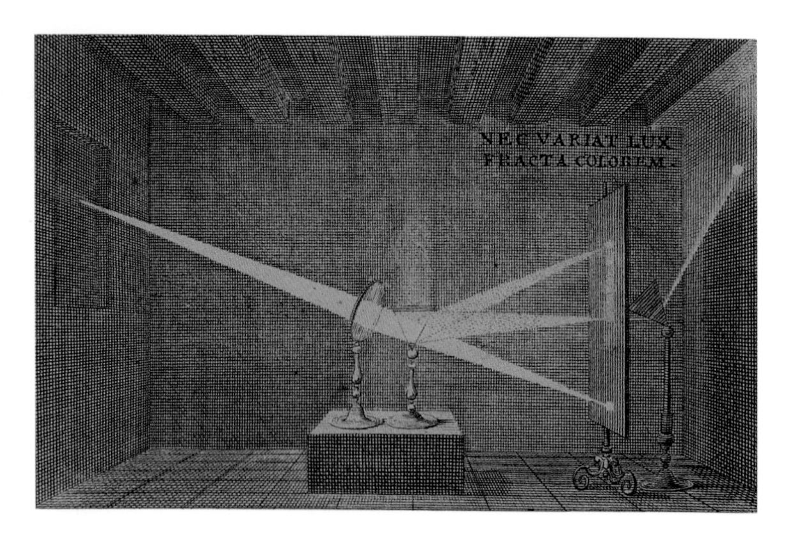

Pierre Varignon oversaw the publication of the second French edition of Newton's *Opticks* (Paris, 1722). At his request, Newton drew this rendition of his "crucial experiment," adding a Latin epigraph that summarized his theory: "Light does not change color when it is refracted." – NYPL-SIBL

Eugène d'Allonville, Chevalier de Louville, mathematician Pierre Rémond de Monmort, and chemist Claude-Joseph Geoffroy, as well as the Italian abbé Conti – as fellows of the Royal Society and to instruct the Society's Curator of Experiments, John Theophilus Desaguliers, to demonstrate the verity of his prismatic experiments. To convince the Frenchmen of his priority over Leibniz, they were also treated to an exhibition of Newton's mathematical papers.

Such calculated hospitality toward foreign visitors, which became a common practice of Newton, reaped handsome dividends. The news of the successful London demonstrations inspired new efforts in Paris. Cardinal de Polignac, who served as president of the Académie in 1718, used his term in office to induce Jean Truchet (Père Sébastien) to prepare the experiments, the cost of which the Cardinal himself bore. As Sébastien subsequently informed Newton, the trials he carried out in 1719 confirmed most of Newton's experiments, "in the presence and with the approval of the most eminent Cardinal de Polignac and also that of a very numerous group of aristocrats and men deeply versed in physics, who are, as it were, blood-brothers of yours." Two years later, Polignac was instrumental in arranging for another trial. Among the spectators on this occasion was Henri François d'Aguesseau, Chancellor of France, who took an active role in the translation of the *Opticks* into French, not for the author's sake, he wrote Newton in September 1721, "but for that of all France, who freely and with applause offers her tongue and her speech to you; nor for the sake of France alone, but on behalf of all philosophers and mathematicians of every nation everywhere, who already look up to you and venerate you as obvious master of them all."[14]

Neither these trials, nor the French translation by Pierre Coste – the handsome second Paris edition (1722) included an illustration by Newton that was drawn by him especially for the edition – put an end to the debates over Newton's theory of light and colors. However, future controversies rarely involved members of the Académie. Although not all academicians were unconditional converts to Newtonian theories – in fact, many of them maintained a concurrent commitment to vortices – their acquiescence indicates the possibility of simultaneously embracing elements of Newtonian science and Cartesian science. And Newton did have his share of committed proponents. As early as 1714, the Oratorian Charles-René Reyneau wrote admiringly of the *Principia*, which he had read in its first edition several years earlier – indicating at least partial conversion. The astronomer Joseph-Nicolas Delisle was an outspoken defender of Newton for at least a decade before leaving France in 1724 to take up a position at the St. Petersburg Academy of Sciences. His friend Chevalier de Louville had published in 1720 in the *Memoirs* of the Académie a paper on the earth's orbit, remarkable as much for its approbatory exposition of universal gravitation as for its failure

This dazzling representation of the Cartesian vision of the solar system – a plate from Bernard de Fontenelle's *Entretiens sur la pluralité des mondes* (Amsterdam, 1719) – depicts, at center, the planets and their satellites as they swirl around the sun engulfed in their respective vortices. Other solar systems in turn swirl around our solar system. – NYPL–SIBL

to mention Newton by name, opting for Kepler's instead. Undoubtedly, Louville engaged in deliberate subterfuge, his motivation discretion – the same motive that informed the tactful manner in which he rejected Cartesian vortices.[15]

Louville's taciturnity was not unlike that exhibited by Varignon vis-à-vis the issue of central forces. Like Varignon, Louville, too, found his papers subject to Fontenelle's editorial censorship whenever his ideas seemed to deviate from the line toed by the Secretary.[16] Ostensibly, Fontenelle expurgated polemics – thus adhering to the Académie's official stance to steer clear of controversies – or sought to render the *Memoirs* more accessible to the wider public. In practice, however, Fontenelle did not hesitate to exercise his authority against those critical of Cartesian ideas, right up until he resigned from his position as Secretary in 1740. Nor did he ever waver when it came to his core Cartesianism. In 1752, at age ninety-five, he published his *Théorie des Tourbillons Cartésiens avec des réflexions sur l'attraction* ("Theory of Cartesian Vortices with Reflections on Gravity"), wherein he defiantly delineated a worldview commensurate with that informing his *Entretiens sur la pluralité des mondes* ("Conversations on the Plurality of Worlds"), published some six and a half decades earlier.

Understandably, then, outsiders could draw exaggerated conclusions regarding the monolithic nature of French science and philosophy. A case in point is the Scottish man of letters domiciled in Paris, Andrew Michael Ramsay, who commented in 1729 that "the French philosophers at present chiefly follow Malebranche. They admire Sir Isaac Newton very much, but don't yet allow of his great principle. 'Tis his particular reasonings, experiments and penetration for which they so much admire him."[17] Even Fontenelle conceded the widespread admiration of Newton on the Continent in his 1728 *éloge* of the English natural philosopher – the significant exception to his usual belittlement of Newton's ideas. There, the Secretary provided not only an accurate and positive account of Newton's discoveries, but inaugurated the fashion of shrewdly (and pleasingly) contrasting Descartes and Newton:

These two great men, who are so strangely opposed to each other, had been closely alike. Both were geniuses of the first order, born to dominate other minds, and to build empires. Both being excellent geometers they saw the need to import geometry into physics. Both founded their physics on geometry which their intellects had framed. But one of them, flying high, sought to take his place at the head of everything, to master first principles by means of a few clear, fundamental ideas in order to descend thereafter to the level of natural phenomena as their necessary consequences. The other, less bold or more modest, set about his business by relying upon phenomena in order to rise to unknown principles, resolved to accept them to the extent that they followed from the order of things. The former starts from what he clearly understands to find the cause of what he perceives, the latter starts from what he perceives to discover its cause, whether clear or obscure. The former's evident principles do not always lead him to phenomena as they are, the phenomena do not always lead the other to evident principles.[18]

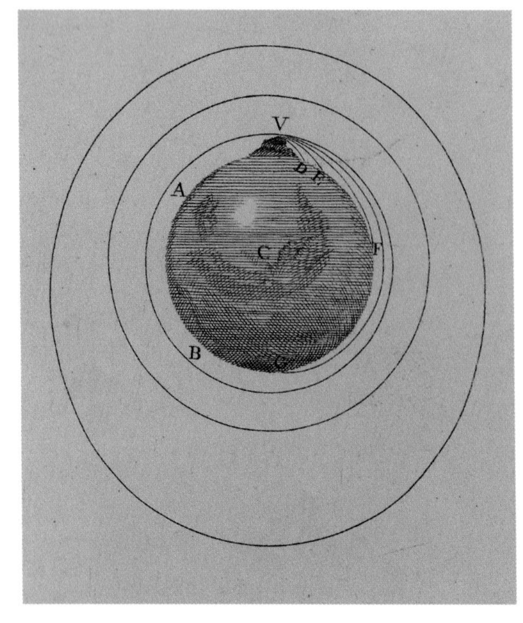

Already in the 1680s Newton argued that if a projectile were to be fired at a sufficiently great velocity, it would enter into an orbit around the earth. This illustration was drawn by Newton for *A Treatise of the System of the World*, an abortive effort to popularize Book III of the *Principia*. It was first published only in 1728. – NYPL–SIBL

The Difficulty of the *Principia* and Its Transmission to a New Generation

In his *éloge*, Fontenelle also raised the issue of the *Principia*'s incomprehensibility, and it is worth pausing on this topic – crucial for the understanding of the reception of Newton's ideas. As noted in chapter 2 above, the issue was a subject of consternation among early English readers of the *Principia*, and by the early 1690s had become a source of witticisms: a Cambridge undergraduate, spotting Newton taking a stroll, reportedly quipped, "here goes a man who wrote a book neither he nor any body else can understand." As if to confute such a perception, in 1691 Newton asserted to a friend eager to master the *Principia* that a comprehension of Euclid's *Elements*, a good text on conic sections, van Schooten's edition of Descartes' *Geometry*, and a treatise on Copernican astronomy were "sufficient for understanding my book" – though, he added, grounding in Christiaan Huygens's *Horologium oscillatorium* would "make you much more ready." Only years later did he concede that such a modest reading list was inadequate preparation.

Incomprehensibility, however, was not simply the inevitable by-product of the need to master a new language of mathematics or to assimilate that mysterious concept of action at a distance (i.e., the action of force between two bodies across empty space without contact). The theory and structure of the *Principia* challenged – and often defeated – even the very few who possessed the requisite mathematical skills, if not by its recondite nature, then by Newton's language and his failure to supply proofs to various propositions. D. T. Whiteside, the editor of Newton's mathematical papers, instanced several examples when even Huygens and Leibniz found it difficult to "appreciate the underlying subtleties of the *Principia*'s dynamical basis on first reading it." A case in point: When confronted with proposition 41 of Book I – "Supposing a centripetal force of any kind and granting the quadratures of curvilinear figures, it is required to find the trajectories in which bodies will move and also the times of their motions in the trajectories found" – Continental mathematicians were unable to "appreciate the depth and power" of Newton's construction, "or to acknowledge the ease with which it might be applied to the particular cases of the inverse-square and inverse-cube orbits." Similarly, with the exception of Huygens, contemporaries greeted proposition 35 of Book II – "If a rare medium consists of minimally small equal particles that are at rest and arranged freely at equal distances from one another, it is required to find the resistance encountered by a sphere moving forward uniformly in this medium" – with "near-total incomprehension." David Gregory, who was also defeated in his efforts to retrace Newton's calculation, became enlightened only after consulting Newton in person.[19]

Exasperation became proverbial. In 1699, a Scottish visitor to Paris heard Malebranche "mightily" commend Newton, with the proviso "that there were many things in his book that passed the bounds of his penetration," and as late as 1731, Johann Bernoulli gave voice to his own exasperation in trying to penetrate Newton's theory about the shape of the earth: "I tried to understand it. I read and reread what he had to say concerning the subject, but ... I could not understand a thing. I do not know whether the fault lies with my impatience, resulting from my reaction in his references to things back in Book I, or whether I do not understand how he applied these [to his theory of the earth's shape in Book III]. In a word, all I found was obscurity and impenetrability."[20]

The seeming incomprehensibility of the *Principia*, then, was rooted in its dense proliferation of radical ideas, ambiguities, and geometrical constructions that neither followed traditional canons of mathematical intelligibility nor quite corresponded with those developed by Leibniz and his disciples. Yet it was precisely these "gaps, inconsistencies, and errors" that in all likelihood prompted "researchers to investigate problems that they might not have otherwise gotten around to dealing with quickly."[21] More problematically, however, the failure of contemporaries – many of whom doubted Newton's ability to manipulate the differential calculus as well as they did – to gauge the depth and subtlety of the *Principia* often led them to hastily reject specific Newtonian results or theories or, alternatively, to take credit for results that Newton had already achieved. Add to this the more fundamental problem that for those like Huygens, raised on the philosophy of Descartes, "a complete, intelligible physical explanation required an account of the mechanical cause of the phenomena," while for Newton "phenomenological or descriptive explanations" sufficed.[22] In other words, a lengthy process of assimilation of new language and new concepts was required before conversion to Newtonian ideas was possible.

In addition, difficulty verging on incomprehensibility made transmission to a new generation a formidable task. The predicament of educators was articulated early in the seventeenth century by a popular author of textbooks, who argued that it was "better to teach methodically ordered traditional positions, even if erroneous or questionable, rather than as-yet unmethodized new theories, even if

This idealized rendering of "Descartes Composing His System of the World" was published in Paris in 1791. – Burndy Library

true."[23] Descartes sought to circumvent precisely this impediment by furnishing not only a rigorous method for the comprehension of his "clear and distinct ideas," but also by structuring his *Principles of Philosophy* pretty much as a textbook. His disciples added their own manuals and were thus on the way to consolidating their grip on natural philosophy instruction by the time the *Principia* appeared. Incumbent on the Newtonians, therefore, was the need to explore pedagogical means to dethrone the new orthodoxy.

Nowhere was this need felt more acutely than in Cambridge. William Whiston, Newton's successor to the Lucasian professorship, recalled how in the early 1690s "[we] poor Wretches, were ignominiously studying the fictitious Hypotheses" of Descartes' philosophy, "alone in Vogue with us at that Time." Yet when, in 1697, the young Samuel Clarke asked Whiston whether he [Clarke] should undertake a translation into Latin of Jacques Rohault's extremely popular *Traité de physique* ("Treatise on Physics"), Whiston encouraged the project on the grounds that "the youth of the university must have, at present, some System of Natural Philosophy for their studies and exercises; and since the true system of Sir Isaac Newton's was not yet made easy enough for the purpose, it is not improper, for their sakes, yet to translate and use the system of Rohault ... but that as soon as Sir Isaac Newton's Philosophy came to be better known, that only ought to be taught, and the other dropped." Clarke proceeded with the translation of the Cartesian textbook, but starting with the second (1702) edition, it became a Trojan horse of sorts, the editor adding notes with a Newtonian content that subverted the meaning of the original.[24] This method of infusing new wine into old vessels, though useful as a stop-gap measure, could not ensure the diffusion of Newtonian ideas beyond the British Isles. Rather, it was the conversion of key Dutch universities – the most popular institutions of higher learning in Europe – that made Newtonian proselytizing efforts a success.

The Teachers of Nations

Holland and England fought three wars during the third quarter of the seventeenth century. But in 1688, following the secret invitation to the Dutch Protestant Stadtholder, William Henry, Prince of Orange, to invade England and depose his Catholic father-in-law, James II, the two countries became, in a manner, united. News of the invasion frightened James into exile, causing him to relinquish his throne to William and Mary. While the new order did not stamp out the enmity and commercial rivalry that had engendered past conflicts, it pitted the two nations against the expansionist policy of France's Louis XIV. More to the point, the new order forged a common cultural front, most notably in the domain of science, heralding a succession of influential professors of natural philosophy and medicine who became the earliest, and most effective, propagators of the Newtonian gospel. By virtue of their celebrated institutions of higher learning, the Dutch rose to prominence on the strength of their professorial body, attracting the most cosmopolitan body of students anywhere.

The person to put Holland on the Newtonian map, so to speak, was Burchard de Volder, who received his medical degree in Utrecht in 1669 and was appointed lecturer in logic and natural philosophy in Leiden the following year. De Volder traveled to England in 1674 and visited Newton at Cambridge. In 1687 he received a presentation copy of the *Principia*; when thanking Newton for that "most splendid offspring of your intellect," de Volder recalled with fondness the kindness shown him by the Lucasian Professor a dozen years earlier. As happened with other early readers, the *Principia* left an indelible impression on de Volder. His biographer, Jean Le Clerc, recounts that de Volder read the book "with great zeal and excitement"; in fact, when Huygens complained to de Volder that he "found Newton's idea obscure," de Volder supposedly retorted: "Newton was, indeed, hard and obscure," but "nevertheless his principles were true."[25]

A veteran of the early 1670s Leiden debates over Cartesianism, de Volder has been portrayed as a Cartesian – and so he undoubtedly was for much of his career. Yet his commitment to Cartesianism was never rigid. Nor did it prevent him from simultaneously pursuing experimentalism. During his 1674 visit to England, de Volder was also inspired by the researches of Robert Boyle and members of the Royal Society, and upon his return to Holland he convinced the Leiden Curators to lay out the considerable funds necessary to purchase scientific instruments and establish a *Theatrum physicum* (physical laboratory). For the next three decades, de Volder proceeded to teach mathematics, Cartesian physics, and experimental natural philosophy in evident harmony. With uncanny ease, he dissented from Descartes on certain key issues as well as eschewed hypotheses in physics. Likewise, he found nothing problematic about incorporating into his lectures results from England, notwithstanding his recourse to Rohault's Cartesian *Traité de physique* as his textbook. Nor should it come as any surprise that before 1687, de Volder's public pronouncements were Cartesian in nature. What alternative was there? In the course of the 1690s, in contrast, his ardor for the Cartesian position cooled appreciably. De Volder still could be counted upon to vigorously defend Descartes in 1696, but this he did in response to a particularly hostile, and influential, critique of Cartesianism by the celebrated scholar Pierre-Daniel Huet.

Jacques Rohault (1620–1675) was one of the most influential second-generation Cartesians, and his *Traité de physique* ("Treatise on Physics") became a standard Cartesian textbook. Newton's friend Samuel Clarke translated the book into Latin, and successive editions included Newtonian notes that subverted the meaning of the text. – Burndy Library

De Volder's change of heart was made public in 1698 when he stepped down as Rector Magnificus. In an oration entitled "On the Power of Reason" (*Oratio de rationis viribus*), de Volder expressed his admiration for the physico-mathematical method developed by Huygens and Newton, contrasting it with what he now considered to be the misguided attempt to apply reason to physics. He confirmed his earlier confidence in the indispensability of mathematics to physics and, as if carping on Newton's own sentiments, warned: "in physics, no matter how certainly we may draw conclusions from an hypothesis, it remains uncertain whether the bodies we have assumed in our reasoning truly exist or not." De Volder proceeded to recommend the application of mathematics and mechanics to other areas, such as medicine: "physicians should study the human body in the same way as Newton studied the universe," he exhorted; "more attention should be given to fluids, which must be studied according to the laws of mechanics."

De Volder retired from his professorship in 1705. He blamed illness but, in Le Clerc's view, just as important was de Volder's "weariness with teaching Cartesian physics and metaphysics and his reluctance to begin building a new system." Certainly, his increasing disillusionment had been evident for at least a decade. After visiting him in 1698, Johann Bernoulli wrote Leibniz of de Volder's estimation that "the Cartesian principles [are] inadequate and even largely incorrect." In view of the large number of students who flocked to de Volder's lectures, Bernoulli urged Leibniz to recruit the Leiden professor to his camp and thereby assure the propagation of his philosophy. Leibniz followed the advice, but failed to win a new convert.[26]

Archibald Pitcairn (1652–1713) was an early proponent of Newtonianism who effectively utilized both Newtonian methodology and matter theory in order to establish medicine on what he believed to be the certain foundation of mathematical physics. – By permission of the Royal College of Physicians of Edinburgh

De Volder's increasing disillusionment with Descartes, and his conviction that medicine, too, should be modeled on mathematics and mechanics, may well have been heightened by the brief tenure of the Scot Archibald Pitcairn at Leiden. Pitcairn received his M.A. degree from the University of Edinburgh in 1671, and subsequently pursued mathematical and medical studies, becoming an early convert to Newtonian ideas. In November 1691, he was appointed to the Chair of the Practice of Medicine in Leiden. On his way to assume the position, Pitcairn paid a visit to Newton and received a copy of the latter's "De natura acidorum" ("On the Nature of Acids"). This important précis of Newton's (private) theory of matter was grounded on Boyle's corpuscularian chemistry as well as on elements of the older chemistry of substances, which Newton rooted in his own concept of short-range attractive force – analogous, but not identical, to gravity. In his inaugural oration, Pitcairn explicated his "iatromathematics" – a holistic approach to the life sciences, which conjoined mathematics, mechanics, and medicine as the sole means to comprehend physiology. "It behooves Physicians to follow rather the Example of Astronomers," Pitcairn admonished, than the "conjectures" and "dreams" of philosophers. Without mentioning either Descartes or Newton by name, Pitcairn pointedly rejected most elements of the former's philosophy – subtle matter, "Fear of a Vacuum," and systems more generally – while advocating Newton's modest methodology of deducing reasons from the phenomena of nature through experiments and mathematics.[27] Pitcairn remained in Leiden for only a year, but his lectures exerted considerable influence on, among others, the young Hermann Boerhaave.

Boerhaave was intended for the ministry, but after attending de Volder's lectures in 1689, his interest shifted gradually to natural philosophy and medicine. In the following decade he practiced medicine and taught mathematics privately until his appointment in 1701 as lecturer in medicine at Leiden. In 1709 he was appointed professor of medicine and botany, adding professor of chemistry in 1718. Almost overnight Boerhaave became the most celebrated professor of medicine and the life sciences in Europe, and in the course

Hermann Boerhaave (1668–1738) was probably the most celebrated professor of medicine (and related disciplines) in the first half of the eighteenth century. In his lectures at Leiden University and his publications, he advocated the application of Newtonian methodology and principles to the life sciences and chemistry. – Smithsonian Institution Libraries, Washington, D.C.

of his thirty-five-year tenure raised two or three generations of physicians and researchers of the life sciences.

From the outset of his teaching career, Boerhaave made his opposition to Cartesianism clear. In 1703, while strongly advocating the introduction of "true" mechanics and mathematics to medicine, he denounced Cartesian mechanists, who attempted "to govern the human body by precepts derived from fictitious principles." In 1709 he again denounced the futile conclusions of the Cartesians deduced "from fictitious causes" and reaffirmed his confidence that the application of mechanics to medicine – he cites Pitcairn approvingly – would assure the simplicity and verity of the discipline. Descartes himself is rebuked a few years later for trying "to deduce from a few principles a wholly improbable world picture" instead of patiently gathering facts and availing himself of mathematics. In contrast, Boerhaave embraced British philosophy, which more and more he equated with Newton.

In 1715, on the occasion of his appointment as Rector Magnificus, Boerhaave delivered his *Sermo Academicus de comprando certo in physicis* ("Academic Discourse on the Achievement of Certainty in Physics"), wherein he gave public expression to his wholehearted endorsement of Newton's reorientation of natural philosophy. Atoms, the void, and weight, Boerhaave noted, had been despised for as long as Descartes' authority prevailed, but at long last they were "vindicated by the Prince of Philosophers, Isaac Newton," and successfully reinstated by means of "wholly irrefutable arguments deriving from the very depth of mathematics." Boerhaave was not bothered by man's inability to grasp the essence of atoms; knowledge of what experiments could teach about their properties was adequate. The same methodology held true for universal gravitation – which Boerhaave considered "absolutely true," no matter that it eluded comprehension; though the origin and

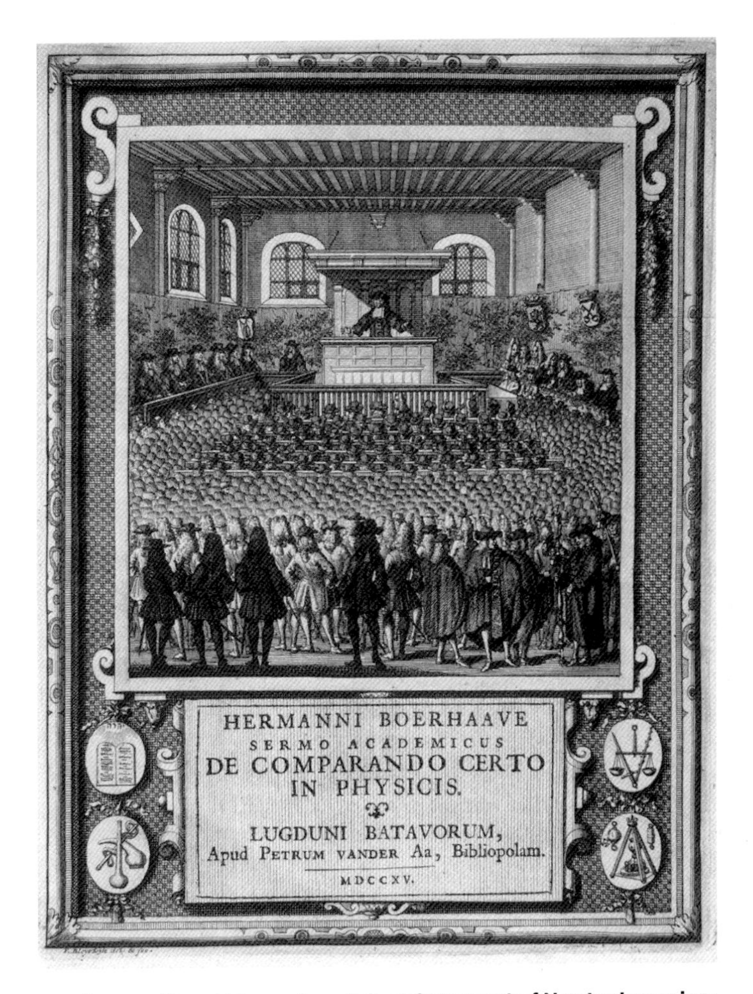

Boerhaave offered his most explicit endorsement of Newton's reorientation of natural philosophy in the inaugural address he delivered upon assuming the office of Rector Magnificus of Leiden University in 1715. – Burndy Library

nature of gravity could not be explained "from mechanical principles," its effects could be "excellently elucidated." And that was what mattered. Boerhaave proceeded to recommend that his auditors embrace Newton's modesty and apply his teaching to their studies; unlike Descartes and Huygens, he stressed, Newton never strayed beyond the bounds of mathematics.[28]

Adding the professorship of chemistry to his other duties in 1718, Boerhaave delivered his "Sermo Academicus de chemia suos errores expurgante" ("Discourse on Chemistry Purging Itself of Its Own Errors)." After recounting for

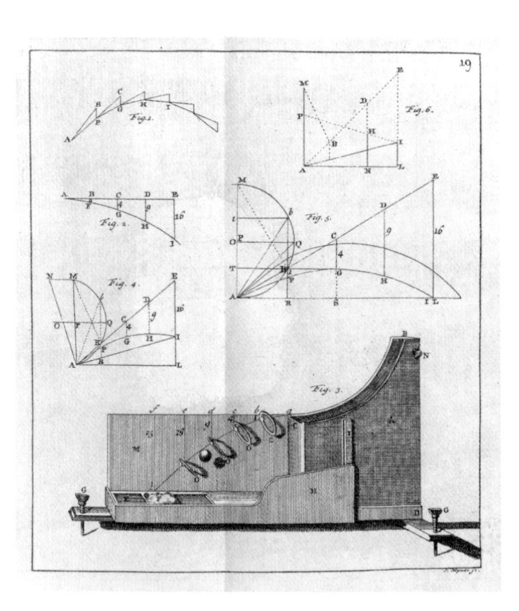

Willem Jacob van 's Gravesande (1688–1742) was professor of mathematics and philosophy at Leiden from 1717 until his death. His lectures and publications – including the Leiden, 1746 and London, 1747 editions of *Elemens de physique démontrez mathématiquement*, from which these illustrations are drawn – rendered Newton's abstruse mathematical physics into visual and dramatic experimental demonstrations. –Portrait: NYPL–Print Collection; plates: NYPL–SIBL

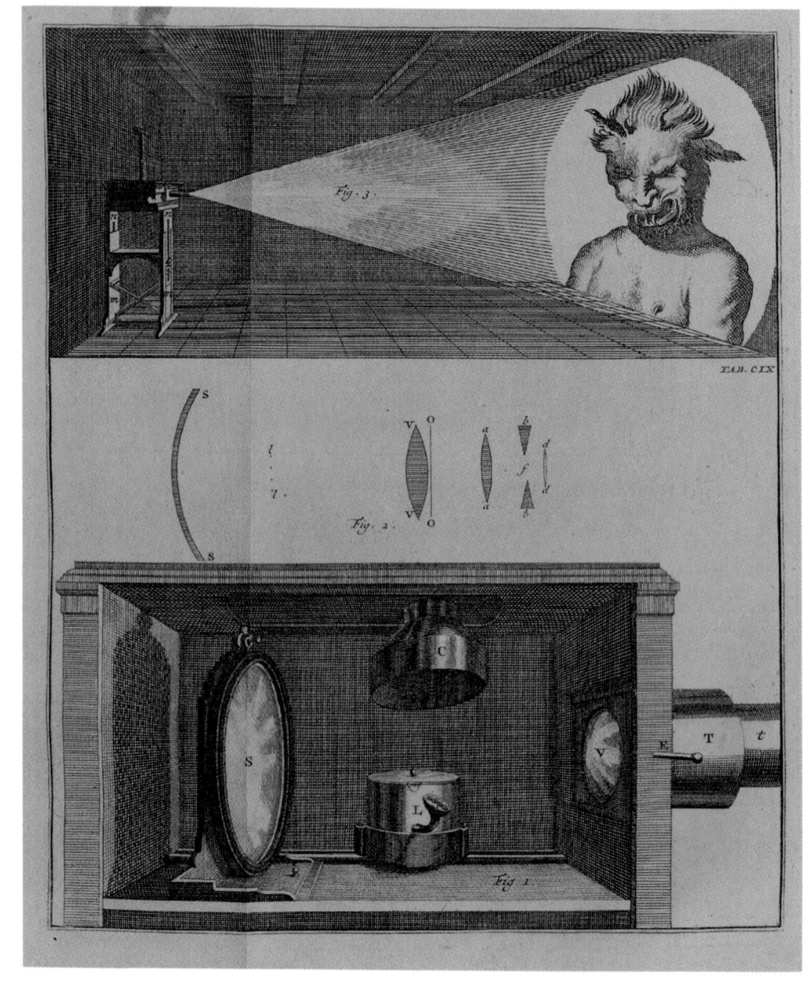

nearly an hour the nature and goals of the discipline, the professor reached a crescendo

> by naming that man in whom Nature has revealed the acme of human perspicacity: Isaac Newton. His achievements are such that during his lifetime no sane person will query his supremacy in philosophy. And when he explains the laws, actions, and forces of bodies - basing himself upon the careful study of their effects - he appeals to chemistry and to nothing else; when he again relates the forces, so found, to other phenomena that are still to be explained he calls upon purely chemical methods; and through his illustrious example he demonstrates that if chemistry did not exist it would be impossible for even the most perspicacious of mortals to gain insight into the proper nature and forces of single bodies.[29]

Even making allowance for the rhetorical flourish mandated by such festive occasions, Boerhaave's programmatic pronouncements leave little doubt of the broad Newtonian outlines of his teaching. Julien Offray de la Mettrie, who studied with Boerhaave in the early 1730s, left testimony that corroborates his mentor's orientation:

> Boerhaave was a Newtonian, convinced and convincing; I saw the most zealot Cartesians yield, despite themselves, to the strength of his demonstrations; he regarded Descartes as a man drunk with fancy and imagination, who had allowed himself to be carried away impetuously with the itch to build a system without consulting Nature; and Newton as the favorite of Nature, as the organ which it used to illuminate the universe, and to him reveal its mysteries that had been well beyond the limits of human understanding; as a man who set out only according to experiment, and established nothing that was not on the most solid foundation.[30]

His zeal notwithstanding, Boerhaave was a professor of medicine, not in charge of the standard courses on mathematics and natural philosophy. De Volder's successor, Jacques Bernard, was interested primarily in ethics and pneumatics (though certain disputations over which he presided suggest his leaning toward Newtonian ideas).

Only in 1717 did the Leiden curators appoint a professor capable of teaching mathematics as well as theoretical - and experimental - natural philosophy: Willem Jacob 's Gravesande. Like his colleague Boerhaave, 's Gravesande was raised to another profession - in his case law - and he actually practiced law after graduating in 1707, cultivating mathematics in his spare time. In 1713 he co-founded the *Journal Litéraire de la Haye*, which immediately took Newton's side in the calculus priority dispute. Traveling to London in 1715 as part of the Dutch delegation to congratulate George I on his ascension to the throne, 's Gravesande remained in England for more than a year, befriending Newton and many members of the Royal Society, and refining his experimental skills through collaboration with Desaguliers. In no small part thanks to Newton's recommendation, 's Gravesande became professor of mathematics and astronomy at Leiden in 1717. In a letter thanking Newton for his gift of the second edition of the *Opticks*, he both reiterated his gratitude and expressed his conviction that Newton's "way of philosophizing" would "be more and more followed in this country." He further flattered himself for having "had some success in giving a taste" of Newton's philosophy in Leiden. In his inaugural address the previous year, 's Gravesande had already articulated his Newtonian profession of faith, hailing Newton as the "restorer of the true philosophy," the first to reject hypotheses in physics, and the person to prove that mathematics alone could serve as the foundation of physics.

In his 1718 letter to Newton, 's Gravesande revealed that "as I talk to people who have made very little progress in mathematics

Petrus van Musschenbroek (1692–1761), who succeeded 's Gravesande as professor at Leiden, was as successful as his predecessor in rendering Newtonian philosophy intelligible through experiments and demonstrations. *Essai de physique* (1739) was one of his more popular textbooks. - Burndy Library

I have been obliged to have several machines constructed to convey the force of propositions whose demonstrations they had not understood. By experiment I give a direct proof of the nature of compounded motions, oblique collisions, and the effect of oblique forces and the principal propositions respecting central forces." Herein lay his greatest contribution. Through his public demonstrations and publications, he succeeded in making intelligible many of Newton's ideas regarding motion, forces, and gravity. Centrifugal motion, the collision of bodies, the motion of fluids, as well as magnetism and optics, were all rendered visible, often spectacularly so. 's Gravesande also powerfully propounded the Newtonian dictum of ignoring first principles in favor of extricating from natural phenomena certain laws of nature, including universal gravitation.

But for all his partisanship, 's Gravesande was no slavish disciple. In 1722 he became a convert to the Leibnizian conception of force, primarily on experimental grounds. It is not surprising that Newton's disciples regarded this as a betrayal tantamount to defection to the enemy camp. 's Gravesande responded that he swore by the words of no master, and that he followed Newton where it truly mattered; avoiding hypotheses and adhering to Newton's own precepts, he followed nature as his guide. As far as he was concerned, his worldview had not changed from its earlier articulation in the preface to his *Physices elementa mathematica experimentis confirmata* ("Mathematical Elements of Natural Philosophy, Confirmed by Experiments"; London, 1719): "I shall always glory in treading in their [the English] Footsteps, who with the Prince of Philosophers [Newton] as their Guide, have first opened the way to the Discovery of truth in Philosophical Matters, dismissing all feigned Hypotheses out of Philosophy."[31]

Four years 's Gravesande's junior, Petrus van Musschenbroek is closely linked with his compatriot, both as an experimenter and as

the author of influential scientific textbooks instrumental in the diffusion of Newtonian ideas. The son of a prominent maker of scientific instruments, Petrus had an elder brother, Jan, also a talented mathematician, who took over the family business in 1707, supplying exceptional instruments to experimental natural philosophers throughout Europe. Petrus eulogized his brother rightly in 1748, acknowledging his "great assistance to the famous 's Gravesande, to me and to all other philosophers, and that it is thanks to his work that we have been able somewhat to advance physics."[32] Petrus made Newton's acquaintance in 1717 when he visited England. Upon his return to Leiden, he attended 's Gravesande's lectures and, shortly after receiving his doctorate in philosophy in 1719, was appointed professor of mathematics and philosophy at

Duisburg. He returned to Holland in 1723 as professor of mathematics and philosophy at Utrecht; in 1740 he was summoned to Leiden as professor of both disciplines.

In his inaugural lecture at Utrecht, Musschenbroek referred to Newton as the "highest of mortals." Three years later, upon sending to the "wisest man to whom this Earth has as yet given birth" a copy of his *Epitome elementorum physico-mathematicorum*, the professor boasted of his accomplishments in Newton's service:

> Being an admirer of your wisdom and philosophical teaching ... I thought it no error to follow in your footsteps (though far behind), in embracing and propagating the Newtonian philosophy. I began to do so in two universities [Duisburg and Utrecht] where the trifling of Cartesianism flourished, and met with success, so that there is hope that the Newtonian philosophy will be seen as true in the greater part of Holland.

Musschenbroek was every bit as creative and successful an experimenter as 's Gravesande, and even more prolific as an author of influential textbooks. Likewise, Musschenbroek displayed the same critical spirit and determination to follow Newton's methodology – rather than his authority – in the search for truth. Such traits did not diminish his devotion to the Newtonian cause. But the manner in which both he and 's Gravesande contributed to the dissemination of Newton's mechanics and optics also attests to the dynamic nature of Newtonianism in the face of new discoveries and theories.[33]

Diffusion in Italy

The diffusion of Newtonian ideas assumed somewhat different contours in Italy than in France, in no small part because of the debilitating effects on Italian science of the Roman Catholic Church's condemnation of Galileo in 1633. This is not to suggest that with the banning of heliocentrism Italian science came to a halt, or that talented natural philosophers and mathematicians stopped producing important work. Rather, the element of uncertainty introduced into the domain of science following the placing of Copernicus' *De revolutionibus orbium coelestium* ("On the Revolution of the Heavenly Bodies") on the Index of Prohibited Books in 1616 and, even more profoundly, following the trial of Galileo a decade and a half later, destabilized the delicate balance between the Catholic Church and local educational and scientific institutions. Interference by the religious authorities loomed over the scientific

This fresco, which shows the orbit of Halley's comet, depicts the solar system as it was understood in the 1770s. It is from the Meridian Room at the Museo *La Specola* in Padua, where celestial bodies were observed during their transit at the celestial meridian. – Courtesy of Museo *La Specola*, INAF-Astronomical Observatory of Padua, Italy

community, shaping public attitudes and tempering curiosity with fear.

By the second half of the seventeenth century, the atmosphere of insecurity had become rooted. In 1663, certain of Descartes' writings were placed on the Index of Prohibited Books. Eight years later, Carlo Cardinal Barberini admonished the Archbishop of Naples to be vigilant lest further scandal befall religion as a consequence of "the opinions of a certain Renato de Cartes, who in years past had published a new philosophical system revising the ancient opinions of the Greeks on atoms, on the basis of whose doctrine some theologians" were producing pernicious interpretations of transubstantiation. The Cardinal may well have had in mind the bold attempt by the Reader of Logic in Pisa, Donato Rossetti, to discuss atomist physics in public. Rossetti was hounded out of Tuscany, but the repercussions of his stirring up of conservative forces boded ill for the scientific community. By the late 1680s, several Neapolitan savants were arrested and put on trial – the so-called "trial of the atheists" – charged with upholding atomist and related "pernicious" philosophical ideas; their persecution underscored the fragility of the freedom to philosophize in Italy.[34]

The growing intellectual insecurity of the seventeenth century was aggravated by the absence of any consistent Church policy. Prudence and friends in high places were often sufficient to forestall overzealous conservatives from retaliation, but an atmosphere of persistent concern weighed heavily on the psyches of Italian savants. Caution and self-censorship became their *modus vivendi*, further prompting those in charge of secular institutions to avoid controversy at almost all costs. A case in point is Prince Leopold de' Medici, who took extreme measures to ensure not only that his creation, the Accademia del Cimento, followed a carefully delineated experimental track – befitting its name, "Academy of the Experimenters" – but also that the published volume summing up the Cimento's decade-long (1657–67) activities confined itself to describing, without any theoretical or even explanatory framework, some 250 experiments. In 1691, Grand Duke Cosimo III proved more cautious still; he prohibited the teaching of atomism at the University of Pisa, stipulating that Aristotle's philosophy alone was to be followed. Along similar lines, the projected statutes of the Academy of Science in Bologna a decade later included a clause enjoining its members to "promise as much for astronomy as for any physical principle in experimental philosophy ... to bring everything into conformity with the Holy Roman Church."[35] The clause, dropped from the final version of the statutes, is nevertheless indicative of the prevailing timidity.

For the many mathematicians and natural philosophers who were priests, and Jesuits in particular, the new state of affairs was particularly difficult: to their vows of obedience were now added additional prohibitions imposed by the various religious orders. In 1651, for example, the general Congregation of the Jesuit Order proscribed a large number of philosophical and scientific propositions pertaining to, among other things, atomism and matter theory. The action immediately affected the venerable mathematician Orazio Grassi, who was on the verge of publishing a treatise on optics. He decided to forgo publication, Grassi wrote a friend in 1652, "because of the strict regulations issued in the last Congregations general, in which we are prevented to teach many opinions, some of which are the substance of my treatise. They say that such opinions are forbidden not because they are bad and false, but only because they are new and uncommon. It will therefore be convenient for me to sacrifice my opinions to the holy obedience, and I am sure that by acting this way I shall have more to gain than to lose." Similar condemnations of atomist and Cartesian views were issued intermittently during the following decades, limiting the ability of Jesuits to teach – let alone publish – novel and controversial ideas.[36]

The problems facing philosophy professors, however, transcended the actual content of the new science. The need to preserve scholasticism for doctrinal reasons remained acute within Catholic culture, and hence the very focus of, say, Newtonian science could be problematic. In this regard, the case of the young Carlo Benvenuti is instructive. A protégé of the influential Jesuit proponent of Newtonian ideas Roger Boscovich, Benvenuti caused a storm at the Collegio Romano in 1752 through two successive public disputations, which presented a truly modernist natural philosophy. Benvenuti then aggravated matters by publishing the theses as *Synopsis physicae generalis* ("Outline of General Physics"). What offended his superiors, however, was not Benvenuti's substitution of Newtonian explanations for those of Aristotle or Descartes as much as it was his turning natural philosophy into a mathematical and experimental science that omitted almost entirely such traditional topics as ontology, pneumatology, and natural theology – all central to the educational objectives of the Jesuits, who sought to unify physics, metaphysics, and theology.[37]

Jesuit practitioners could demur only in private. Wryly commenting in 1693 on the horrifying (and inhibiting) effects of the "trial of the atheists," a Neapolitan Jesuit confided to Vincenzo Viviani, Galileo's last pupil, that he found himself living "in a country where one does no other experiment than to change skin, hoping that the new skin withstands the sword thrusts of duels." Nearly seven decades later, Boscovich expressed similar sentiments. Notwithstanding his supreme efforts – and his friendship with the enlightened Pope Benedict XIV – he was unable to influence the myopia of those in power, who scarcely differentiated between the new natural philosophy and heresy. "Believe me," he wrote his brother in 1760, "I

The celebrated eighteenth-century Jesuit natural philosopher and poet Roger Joseph Boscovich (1711–1787) was for many years professor of mathematics at the Collegio Romano. – NYPL–Music Division

turn cold at the thought of having to return [to the Collegio Romano], I have lost all my love for that house though you should know that there are many people there who have been good to me. Those who are good don't count and the studies of those who count will come to nothing. There, if you are not a Peripatetic you are a heretic."[38]

The terror of being charged with heresy lay heavily on all philosophers and men of letters, irrespective of whether they were priests or not. Few, if any, sought martyrdom. Circumspection became the order of the day. "Learn at Galileo's expense," the mathematician and future cardinal Michelangelo Ricci wrote a correspondent in 1658. "He ran into so much trouble just because he picked fights."[39] Ricci also recognized the virtues of a quiet campaign – as evidenced in his attempt to rehabilitate Galileo during the second half of the seventeenth century – a course of action the fiery Tuscan would not have stomached a generation earlier. And prudence begot dissimulation among those seeking to publish. Cesare Beccaria, the celebrated author of *Dei delitti e delle pene* ("On Crimes and Punishments"), commented frankly in 1766 on just such a need in a letter to his French translator. In writing his book, Beccaria admitted, he had had in mind the fate of Machiavelli, Galileo, and the historian Pietro Giannone. "I could hear the rattling chains of superstition and the howls of fanaticism stifling the faint moans of truth. It was this that caused me – forced me –

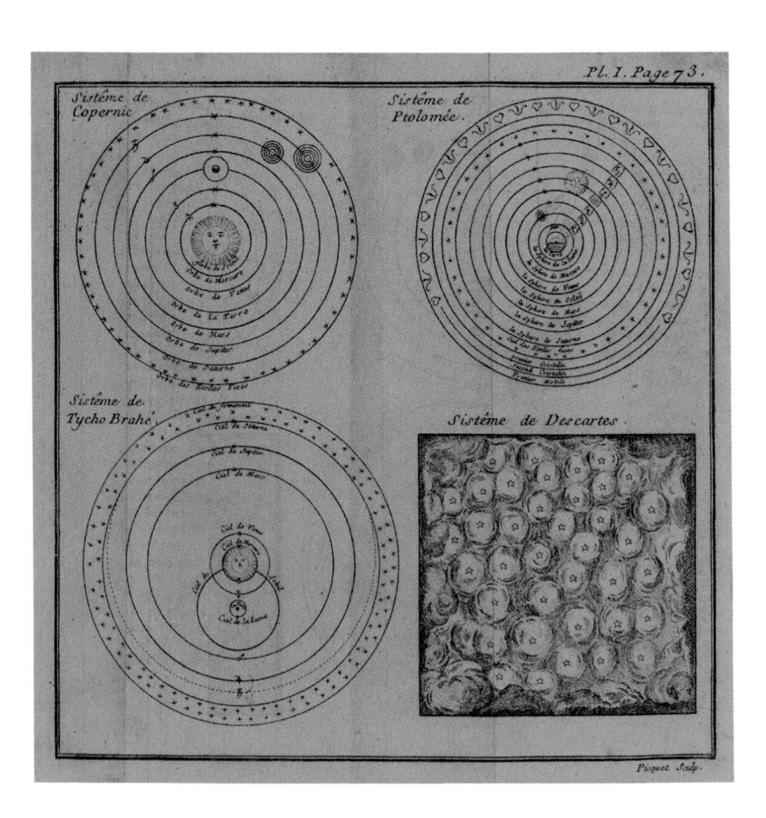

The prohibition of Copernicanism encouraged the adoption of an outwardly noncommittal and eclectic approach by teachers and authors toward the several world systems, although their preferences could be inferred from the manner of presentation. This diagram, displaying the systems of Ptolemy, Copernicus, Tycho Brahe, and Descartes, is characteristic of the practice. From Taitbout, *Abrégé élémentaire d'astronomie, de physique, d'histoire naturelle et de chymie* (Paris, 1777). - NYPL-SIBL

sometimes to veil the light of truth in a pious shroud. I wished to defend humanity without being a martyr to it. The Habitual caution instilled in me by the need to express myself obscurely has sometimes made me do so even when I did not need to."[40]

Such dissembling is in evidence everywhere. Authors routinely couched scientific discussions in probabilistic terms, eschewing along the way the possibility of attaining certitude. The attitude of the Bolognese professor Geminiano Montanari was undoubtedly widespread: "I hold many opinions for probable, many for improbable, and none for absolutely true." On controversial topics, such as cosmology, savants became even more explicit. Giovanni Poleni, the Paduan philosophy professor who will be discussed in detail below, was careful to preface his 1712 discussion of Cartesian vortices and Newtonian gravity in *De vorticibus coelestibus dialogus* ("Dialogue on Celestial Vortices") with an emphatic disclaimer: he considered "the Copernican system completely false," and thus honored "with due veneration the decree with which such a system is justly condemned." Equally emphatic were the two Minim Friars Thomas Le Seur and François Jacquier, the editors of the 1739–42 three-volume edition of the *Principia*. They prefaced Book III with a statement that since Newton took the motion of the earth for granted in this part of the treatise, "any explication of his views must start from the same hypothesis. They must, therefore, put on, as it were, the *persona* of Newton for the occasion. For themselves, they profess their obedience to the papal decree."[41]

Matters were complicated by the fact that natural philosophers and mathematicians, especially university professors, could be charged with transgressions before the religious authorities by anyone eager to discredit them. A correspondent informed the naturalist Francesco Redi in 1685 that ever since the Naples professors of medicine and law began teaching philosophy as well, the regular philosophy professors, whose classes had become empty, retaliated by accusing their opponents of teaching doctrines that were "crazy, foolish, and dangerous to faith." Another philosopher, Giuseppe Roma, proved so effective a teacher of modern physics that an informant of the Roman Inquisition charged him with refusing to "recognize the dangerous things he impresses upon the minds of the young." As for Celestino Galiani, the most zealous proponent of Newtonian ideas in the early eighteenth century, he was convinced in 1713 that had a copy of Samuel Clarke's rendition of Rohault's *Traité de physique* fallen "into the hands of some rogue zealot," he "would have arranged for its prohibition in the Index." Rohault's text escaped such a fate, but Galiani's own appointment in 1731 as *Cappellano maggiore* in Naples elicited denunciations regarding his orthodoxy from those opposed to his reformist ideas. A rumor circulated that Galiani "had persuaded the study of Newton in Naples," to which proselytizing efforts his auditors retorted: "they wished to continue to be Catholics" instead.[42]

Little could be done to escape the religious intimidation that loomed over Italian scientific culture, though initially the Galileans put up a fight. In 1665, for example, Giovanni Alfonso Borelli, a physicist and member of the Accademia del Cimento, asked permission of Prince Leopold de Medici to publish in France his observations on the comet that had appeared in November 1664. Such a course, Borelli reasoned, would enable Italians to gain courage from "the free way of speaking in assemblies of Jesuits and other men of letters" in France – not least about Copernicanism. Consequently, Galileo's condemnation "may become acceptable and less frightening" to Italians as well. Two years later, the mathematician Stefano degli Angeli published a refutation of Giovanni Battista Riccioli's *Almagestum novum* – an imposing folio by a learned Jesuit astronomer who defended the condemnation of Galileo on mathematical as well as theological grounds – purposely, he claimed, in order to salvage the reputation of the Italian scientific community. Careful to declare his adherence to the decree against Copernicanism, degli Angeli nevertheless argued that Riccioli's feeble arguments in favor of the immobility of the earth – as well as his reckless rationalization of Galileo's condemnation – would disgrace the Catholic Church and Italian science abroad.[43]

These and other efforts proved to no avail. A perception regarding the "decline" of Italian science was quick to form and spread. The demise of the Jesuit scientific tradition was articulated in 1686 by the former secretary of the Cimento, Lorenzo Magalotti, who quipped that "a Jesuit mathematician is a rarity worthy of being put into a museum." Four decades later, Antonio Conti – Newton's one-time friend and conduit of his ideas – broadened the perception of decline to include the entire scientific community in the peninsula. From the pages of the *Giornale de' Letterati d'Italia*, "he deplored the impression given abroad that Italian science was no longer important."[44] By the end of the century, a Galilean biographer denounced the Catholic Church not only for its condemnation of Galileo, but for its responsibility for "constraining Italian culture to a perennial state of inferiority in comparison to European science." Outsiders joined the chorus of scoffers. Leibniz commented on several occasions on the scarcity of Italian mathematicians and, analogously, argued in Rome for the need to remove Copernicus from the Index. For his part, Newton reflected on "the happiness of being born in a land of liberty where he could speak his mind, not afraid of Inquisition as Galileo ... not obliged as Des Cartes was to go into a strange country and to say he proved transubstantiation by his philosophy."[45]

This complex religious and political background made the diffusion of Newtonianism

Upon publishing the first edition of his magnum opus, *Principi di una scienza nuova* (Naples, 1725), Giambattista Vico, like his Neapolitan friend Paolo Mattia Doria, sent a copy to Newton, asking his opinion of a work that purported to establish the human sciences on firmer principles than Newton had established for the natural sciences. This frontispiece, from the 1744 Naples edition, depicts Metaphysics standing on a celestial globe, thereby asserting its superiority to natural philosophy. – Harvard College Library

in Italy especially problematic. In addition to the difficulties posed by Newton's recondite and revolutionary ideas – as well as by the Copernican context in which they were couched – local savants had to contend with the hard fact that the English author of the *Principia* and the *Opticks* could be regarded by the religious authorities as a heretic. The perceived nexus between Newton's science and John Locke's epistemology also proved thorny; even before Locke was placed on the Index of Prohibited Books in 1734, his ideas were routinely decried in Italy for their materialist, and atheistic, connotations. Understandably, therefore, much of the study and discussion of Newtonian ideas occurred in private. As a result, to appreciate the scope of the initial diffusion of Newtonianism, it is necessary to scour the correspondence and private papers of Italian savants.

The case of the Neapolitan philosopher Giuseppe Valletta exemplifies such silent diffusion. Having read Christoph Pfautz's favorable review of the *Principia* in the *Acta Eruditorum* in 1688, Valletta promptly ordered a copy. His enthusiasm for Newtonianism began to color his correspondence, but became more evident still in 1709 when he sought out further illumination of the *Principia* from an expert visitor to Naples, the mathematician William Burnet (who would later serve as colonial governor of Massachusetts, New Jersey, and New York). Three years later, Valletta attempted to establish direct contact with Newton. Availing himself of his friend Paolo Mattia Doria's request that he solicit Newton's judgment of Doria's two works *La vita civile* and *L'educazione del principe* ("The Civil Life" and "The Education of the Prince"), Valletta dispatched them to Newton in February 1712, voicing in his cover letter his admiration for the *Principia* and the *Opticks*. He also entreated Anthony Ashley Cooper, Earl of Shaftesbury – then on an extended visit to Naples – to convey his respects to Newton.[46]

Newton appears to have rebuffed both overtures. Yet, even without direct contact, simply by making his copy of the *Principia* available to others, Valletta contributed to the diffusion of Newtonianism in Italy. Agostino Arriani, professor of mathematics at the University of Naples, acquainted himself with Newton's ideas by studying Valletta's copy of the *Principia*, incorporating what he learned into the lectures he delivered in 1701 before the newly established Accademia Palatina del Medinaceli, wherein he extolled the "marvelous" *Principia* and its author as "a man of vast learning and elevated genius." Another beneficiary of Valletta's liberality was the jurist Giacinto de Cristofaro, who turned to the study of mathematics and physics following his condemnation in the "trial of the atheists" in 1697. As he confided to Celestino Galiani in 1709, Valletta's copy had launched his study of the *Principia* years before.[47]

Other early readers of Newton who transmitted their appreciation of the *Principia* to students and friends included the aging Stefano degli Angeli, professor of mathematics at Padua, an early admirer of Newton who put a copy in the hands of his student Jacopo Riccati around 1695. Vincenzo Viviani, the proud recipient of a presentation copy of the *Principia* from Edmond Halley, was undoubtedly the person who kindled the interest of Guido Grandi, who taught mathematics at the monastery of the Camaldolese Order in Florence in the late 1690s. In 1703, Grandi, now professor of philosophy in Pisa, sought to initiate a correspondence with the Englishman, sending him copies of both his own mathematical works and Viviani's. For his part, Domenico Guglielmini had gained familiarity with the *Principia* by the time he published his important treatise on hydraulics, *Della natura de' fiumi* ("On the Nature of Rivers"), in 1697.

Jacopo Riccati (1676–1754) was arguably the greatest Italian mathematician of the first half of the eighteenth century. – NYPL–SIBL

His remarks in the preface make clear the relevance he found in Newton's work to the physico-mathematical project he had undertaken, although his own approach lacked the mathematical rigor of the *Principia* and was more empirical in its methodology. Guglielmini appears to have introduced at least one student, Vittorio Francesco Stancari, to the *Principia*. Certainly, Stancari discussed Newton in lectures he delivered before the Bolognese Accademia degli Inquieti in 1703–4.[48]

The Figures Room at the Museo *La Specola*, Padua, is decorated with life-size frescoes of major figures in astronomy and meteorology, painted in 1772 and 1773 by Giacomo Ciesa. This one depicts Giovanni Poleni (1683–1761), Professor of Mathematics and Experimental Philosophy at the University of Padua. – Courtesy of Museo *La Specola*, INAF-Astronomical Observatory of Padua, Italy

In Padua, the research and teaching of Giovanni Poleni and Jakob Hermann proved crucial for the diffusion of Newtonianism in Italy. Born in 1683, the patrician Poleni received a solid philosophical and scientific grounding from Father Francesco Caro in Venice. Poleni's innate mechanical skills were displayed by his construction of a large Copernican sphere at the close of his course of study in 1703. Five years later, Poleni read Newton's *Opticks*; having tried out several of the experiments, he sought to develop a mechanical explanation for Newton's theory of colors. When Poleni was appointed a lecturer on astronomy and meteorology in Padua in 1709, he voiced his receptivity to Newtonian ideas in his inaugural oration. He would treat astronomy, Poleni declared, by presenting "observable data from which the properties of matter and the forces of nature could be deduced." In discussing comets, he would follow Newton, who "by the admirable eloquence of his sublime genius so diligently emptied the

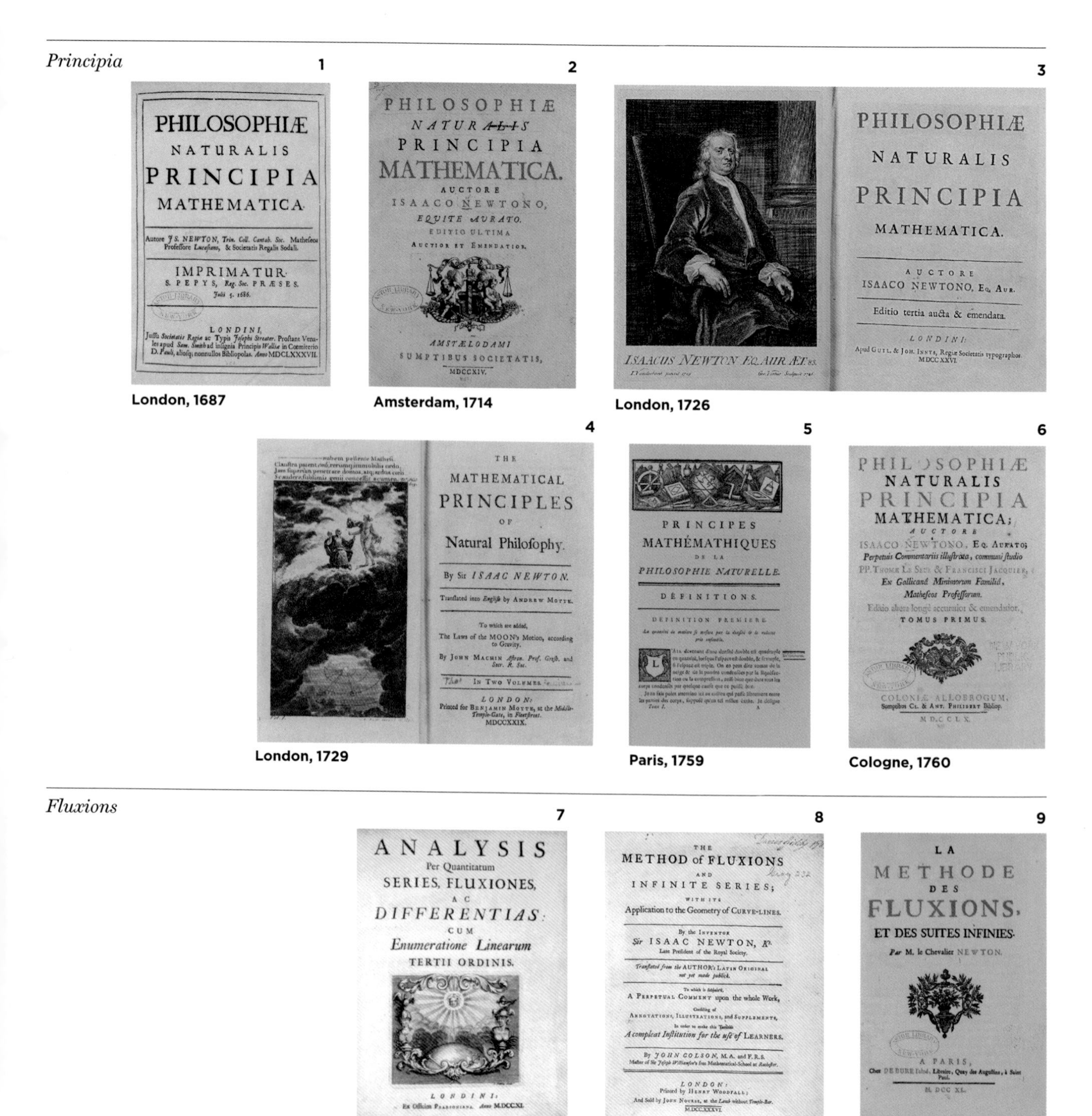

Principia

1 London, 1687

2 Amsterdam, 1714

3 London, 1726

4 London, 1729

5 Paris, 1759

6 Cologne, 1760

Fluxions

7 London, 1711

8 London, 1736

9 Paris, 1740

Opticks

10

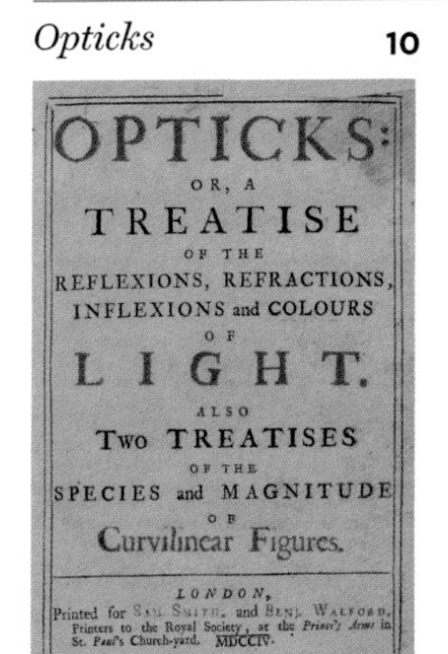

OPTICKS:
OR, A
TREATISE
OF THE
REFLEXIONS, REFRACTIONS,
INFLEXIONS and COLOURS
OF
LIGHT.
ALSO
Two TREATISES
OF THE
SPECIES and MAGNITUDE
OF
Curvilinear Figures.

LONDON,
Printed for SAM. SMITH, and BENJ. WALFORD,
Printers to the Royal Society, at the Prince's Arms in
St. Paul's Church-yard. MDCCIV.

London, 1704

11

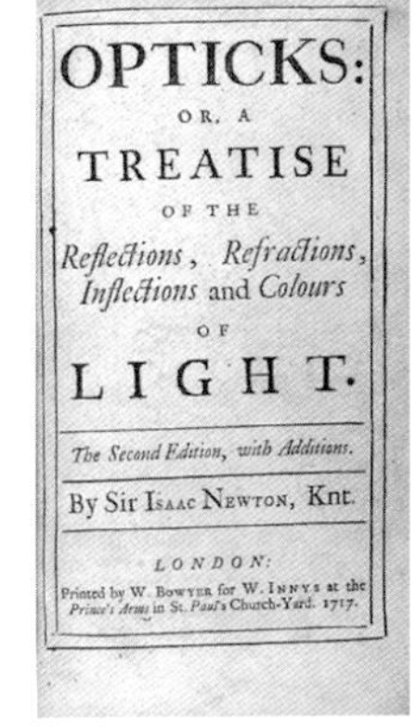

OPTICKS:
OR, A
TREATISE
OF THE
Reflections, Refractions,
Inflections and Colours
OF
LIGHT.

The Second Edition, with Additions.

By Sir ISAAC NEWTON, Knt.

LONDON:
Printed by W. BOWYER for W. INNYS at the
Prince's Arms in St. Paul's Church-Yard. 1717.

London, 1717

12

OPTICE:
SIVE DE
Reflexionibus, Refractionibus,
Inflexionibus & Coloribus
LUCIS,
LIBRI TRES.

Authore ISAACO NEWTON, Equite Aurato.
Latine reddidit Samuel Clarke, S. T. P.
Editio Secunda, auctior.

LONDINI:
Impensis GUL. & JOH. INNYS Regiæ Societatis Typographorum ad Insignia Principis in Area Occidentali D. Pauli. MDCCXIX.

London, 1719

13

TRAITÉ
D'OPTIQUE
SUR
LES REFLEXIONS, REFRACTIONS,
INFLEXIONS, ET LES COULEURS,
DE
LA LUMIERE.
Par Monsieur LE CHEVALIER NEWTON.
Traduit par M. COSTE, sur la seconde Edition Angloise, augmentée par l'Auteur.
SECONDE EDITION FRANCOISE,
beaucoup plus correcte que la premiere.

A PARIS,
Chez MONTALANT, Quay des Augustins, du côté du Pont saint Michel.

M. DCC. XXII.
AVEC APPROBATION ET PRIVILEGE DU ROY.

Paris, 1722

14

OPTICE:
SIVE DE
REFLEXIONIBUS, REFRACTIONIBUS,
INFLEXIONIBUS ET COLORIBUS
LUCIS,
LIBRI TRES.
AUCTORE
ISAACO NEWTON,
EQUITE AURATO.
Latine reddidit
SAMUEL CLARKE, S. T. P.
EDITIO NOVISSIMA.

LAUSANNÆ & GENEVÆ,
Sumpt. MARCI-MICHAELIS BOUSQUET & Sociorum.
MDCCXL.

Lausanne, 1740

15

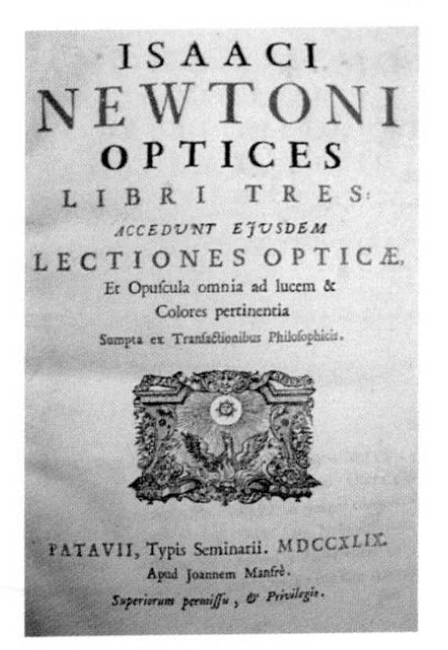

ISAACI
NEWTONI
OPTICES
LIBRI TRES:
ACCEDUNT EJUSDEM
LECTIONES OPTICÆ,
Et Opuscula omnia ad lucem &
Colores pertinentia
Sumpta ex Transactionibus Philosophicis.

PATAVII, Typis Seminarii. MDCCXLIX.
Apud Joannem Manfrè.
Superiorum permissu, & Privilegio.

Padua, 1749

Arithmetica Universalis

16

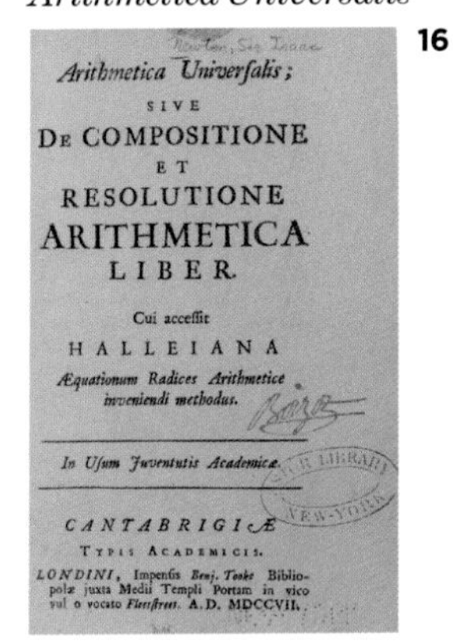

Arithmetica Universalis;
SIVE
DE COMPOSITIONE
ET
RESOLUTIONE
ARITHMETICA
LIBER.

Cui accessit
HALLEIANA
Æquationum Radices Arithmetice inveniendi methodus.

In Usum Juventutis Academicæ.

CANTABRIGIÆ:
TYPIS ACADEMICIS.
LONDINI, Impensis Benj. Tooke Bibliopolæ juxta Medii Templi Portam in vico vulgo vocato Fleetstreet. A.D. MDCCVII.

Cambridge, 1707

Chronology of Ancient Kingdoms

17

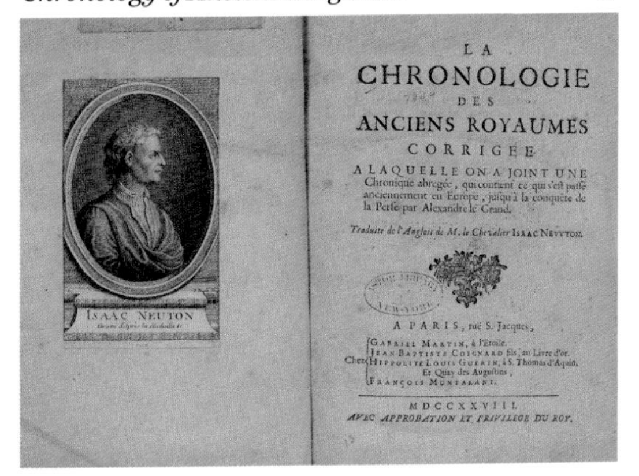

ISAAC NEUTON

LA
CHRONOLOGIE
DES
ANCIENS ROYAUMES
CORRIGÉE.
A LAQUELLE ON A JOINT UNE
Chronique abregée, qui contient ce qui s'est passé anciennement en Europe, jusqu'à la conquête de la Perse par Alexandre le Grand.
Traduite de l'Anglois de M. le Chevalier ISAAC NEWTON.

A PARIS, rue S. Jacques,
Chez GABRIEL MARTIN, à l'Etoile.
JEAN BAPTISTE COIGNARD fils, au Livre d'or.
HIPPOLYTE LOUIS GUERIN, à S. Thomas d'Aquin.
Et Quay des Augustins,
FRANÇOIS MONTALANT.

MDCCXXVIII.
AVEC APPROBATION ET PRIVILEGE DU ROY.

Paris, 1728

Opuscula (Minor Works)

18

ISAACI
NEWTONI,
EQUITIS AURATI,
OPUSCULA
MATHEMATICA, PHILOSOPHICA
ET
PHILOLOGICA.
Collegit partimque Latinè vertit ac recensuit
JOH. CASTILLIONEUS
JURISCONSULTUS.
TOMUS SECUNDUS
Continens
PHILOSOPHICA.

LAUSANNÆ & GENEVÆ,
Apud MARCUM-MICHAELEM BOUSQUET & Socios.
MDCCXLIV.

Lausanne and Geneva, 1744

1, 4: NYPL–Rare Books Division

2, 3, 5, 6, 9, 10, 12–14, 16, 18: NYPL–SIBL

7, 8, 11: Burndy Library

15: Courtesy of the California Institute of Technology Archives

17: NYPL–General Research Division

Giovanni Poleni's ***De vorticibus coelestibus dialogus*** **offered a nuanced account of the efforts by Cartesians to respond to the serious challenge posed by the** ***Principia*** **to the reality of vortices.**
– Burndy Library

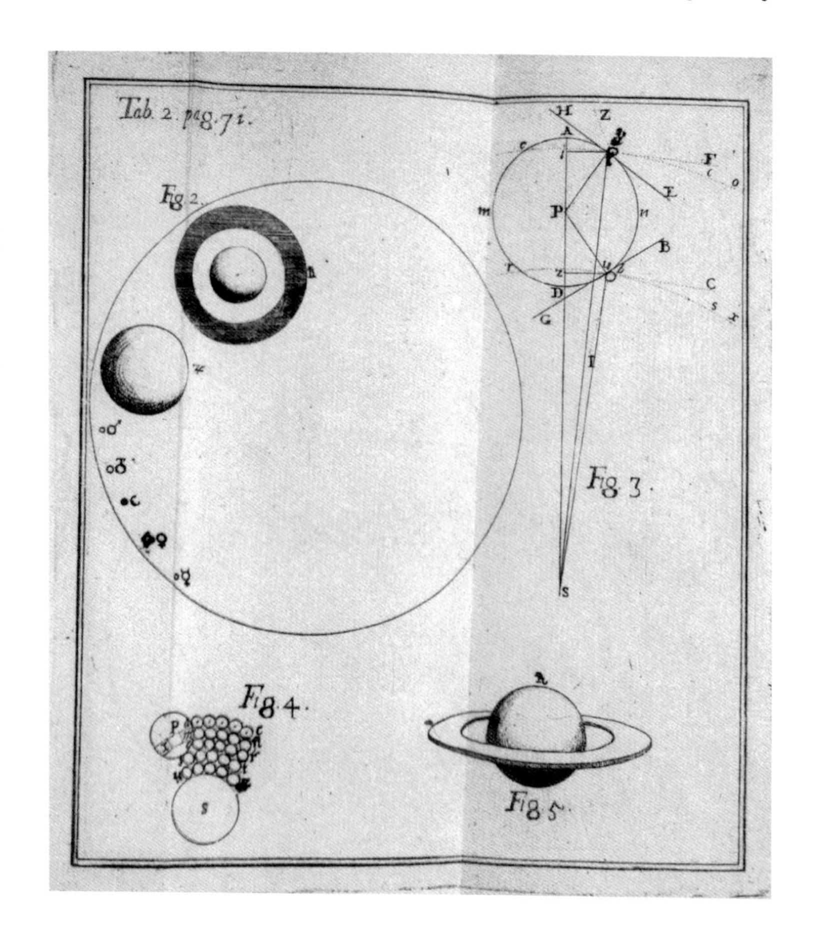

material of the celestial spaces." Such a progressive course, however, would give no occasion for scandal, for Poleni followed the prudence of his predecessors, using such disclaimers as "let us pretend the earth moves" when discussing sensitive matters.[49]

In *De vorticibus coelestibus dialogus*, Poleni presented an informed and well-rounded discussion of the recent, mainly French, attempts to make vortices conform to Kepler's laws – and, by implication, to respond to Newton's powerful challenge to vortices. In fact, Poleni endorsed the cogency of Newton's arguments, although he entertained the possibility that there might exist some sort of aetherial motion that could prove consistent with Kepler's laws. Poleni's guarded approach in *De vorticibus* – which, we noted above, caused him to pronounce at the outset his adherence to a motionless earth – was made explicit in the advertisement devoted to the publication in the *Giornale de' Letterati*: Poleni's aim was neither to establish nor to discredit vortices, but only to "open the path to research the truth, leaving judgment to the learned." Such caution was carried into Poleni's 1716 inaugural lecture upon his appointment as professor of philosophy. The lecture was modeled, to a certain extent, on Hermann Boerhaave's Newtonian *Sermo Academicus de comprando certo in physicis* and, like the Leiden professor a year earlier, Poleni reiterated the conviction that he who builds on experiments "does not risk having to revoke his science as the builder upon untested hypotheses does." To this thinly veiled criticism of Descartes' system, Poleni added his tacit acceptance of the reality of Newton's account of universal gravitation.[50]

Partly by inclination and partly owing to the influence of his colleague Jakob Hermann, by the mid-1710s Poleni had turned away from astronomy to focus on physics and mathematics. Hydraulics became central to his work, and in 1717 he published his important *De motu acquae mixto* ("On the Mixed Motion of Waters"), which relied heavily on Newton for its subject matter as well as for its methodology. For the following four decades, he pursued an active career of teaching and experimentation, thus making him the Paduan counterpart to 's Gravesande and Musschenbroek in Leiden. Though no slavish follower of any master, Poleni's lectures and publications helped initiate numerous young men into Newtonian science. Like 's Gravesande, too, Poleni sided with the Leibnizians in the debates over the measure of forces (the *vis viva* controversy), and for much the same reason: repeated experiments persuaded him of the reality of "living forces." Nevertheless, his admiration of Newtonian science and methodology remained steadfast.

In 1716, Alexander Cunningham, the English envoy to Venice – who was convinced that the *Principia* displayed "such profundity and judgement as far surpassed both the genius and discoveries of antiquity, and the capacity of his own contemporaries" – informed Newton of conversations he had had with Poleni. The Paduan professor expressed his approval of Newton's "Synthetique way in the search of nature," the envoy affirmed, as well as his admiration of Newton's "happyness in finding 'em out." Indeed, the account continued, Poleni had attained such comprehension of Newton's works "as if he had been frequently with you." In a subsequent letter, Cunningham reported that Poleni had informed him that the *Principia* had "inflamed about 20 or 25 of his acquaintances into the studie of Nature and Mathematicks, and that they altogether follow your way." Poleni – who had been elected a Fellow of the Royal Society in 1710 – also accepted Newton's version of the calculus priority dispute, set out in the *Commercium Epistolicum*, judging it "unanswerable"; in contrast, Leibniz's attempt at a rejoinder in the *Charta volans*, Poleni averred, had only confirmed him in his opinion. Such discussions

This allegorical frontispiece appeared in Jakob Hermann's *Phoronomia* (Amsterdam, 1716), which sought to elucidate the principles of the new science of dynamics. – NYPL–SIBL

with Poleni and other Italian savants led Cunningham to exclaim: "I find all that speak of you, which are many, have a true sentiment not only of your sublime learning, but alsoe of your solid judgement and candure, and these they say they draw from your way of writing."[51]

Poleni was not alone in making Padua a gateway for the spread of Newtonianism. The tenure of Jakob Hermann as professor of mathematics there (1707–13) had proven crucial for the simultaneous diffusion of the differential and integral calculus and of Newtonian physics – once again attesting to the important cross-fertilization of the two domains, even in the face of the fierce public disputes between the partisans of Newton and Leibniz. It was, in fact, to Hermann that Poleni owed much of his mastery of the differential calculus and of the *Principia*. Hermann was the most distinguished student of Jakob Bernoulli (Johann's elder brother) in Basel, and in 1700 he published a defense of the differential calculus in response to the criticism of the Dutch mathematician Bernard Nieuwentijt. The following year Hermann embarked on a long tour

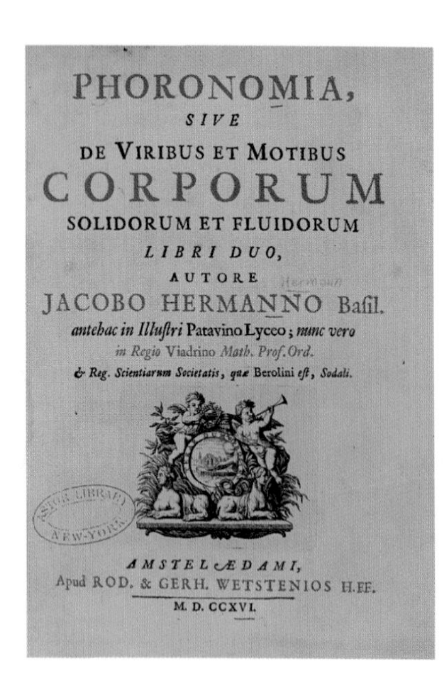
PHORONOMIA,
SIVE
DE VIRIBUS ET MOTIBUS
CORPORUM
SOLIDORUM ET FLUIDORUM
LIBRI DUO,
AUTORE
JACOBO HERMANNO Basil.
antehac in Illustri Patavino Lyceo; *nunc vero*
in Regio Viadrino *Math. Prof. Ord.*
& Reg. Scientiarum Societatis, quæ Berolini *est, Sodali.*
AMSTELÆDAMI,
Apud ROD. & GERH. WETSTENIOS H.FF.
M. D. CCXVI.

In the *Phoronomia*, Jakob Hermannn treated both Newton and Leibniz with the utmost respect. – NYPL-SIBL

that included a stay in England, where he met Newton and several of his disciples. Fifteen years later, upon sending Newton a copy of his *Phoronomia* ("The Theory of Thrust" [Kinematics]), Hermann wrote that although the book was "hardly worthy of [Newton's] glance," he felt compelled to send a copy as testimony of his "undying regard," and his "constant recollection of the kindness and goodwill" he had received from Newton while in London. He concluded: "I justly look up to you as the supreme judge on these matters, and I confess that there are many things in my book, for discovering which the credit belongs to you."[52]

Discussing the disputes over central forces in 1719, the mathematician Brooke Taylor noted that Hermann's contributions to the debate were more respectful of Newton than those of other foreigners – "though in one place he seems to wish that Leibnitz might have a share in his discovery of universal gravitation." Indeed, Hermann always strove to maintain a balance between the two camps. The *Phoronomia*, although dedicated to Leibniz, referred to Newton as the "greatest" geometer and paid tribute to the *Principia* as a "golden book." An introductory poem lavished further praise on the Englishman: "Newton, inhabitant of a wealthy isle, But an island having nothing more gold than himself, Was the first to go along this path; and you perhaps Give nothing of lesser value to the public." For his impartiality, Hermann incurred the displeasure of the Leibnizians. "Hermann is too deferential towards the Englishmen," Johann Bernoulli wrote Leibniz, "to the point that practically all the way through the book he actually uses Newton's style of demonstration."[53] Hermann struck a similar balance in his teaching in Padua – both in his public lectures on mathematics and mechanics and in his private tutorials on the calculus – as well as in the extensive epistolary exchange he carried on with Italian mathematicians and natural philosophers, who turned to him for enlightenment regarding the new language of mathematics and its application to physics.

Hermann's *Phoronomia* consisted of two books, roughly corresponding to the first two books of the *Principia*. Like most Continental mathematicians, Hermann made no effort to extend the discussion of mechanics to the "System of the World" expounded in Book III of the *Principia*. Indeed, an implacable critic, Giuseppe Verzaglia, charged that Hermann's strictly mathematical approach violated the very physico-mathematical method of "the incomparable Newton." Central forces and orbital motions, Verzaglia thundered, could not be treated without taking into account the physical nature of celestial mechanics, which Newton "taught us how to handle ... without ambiguity."[54] Verzaglia was correct, of course, but perhaps it suited the embattled Italian savants not to confront head on the heliocentrism of the *Principia*, at least not initially. The notable exception to the rule can be found in the career of the most committed proponent of Newton in the early eighteenth century, Celestino Galiani.

A monk of the Celestine order, Galiani caught the attention of his superiors, who sent him to study in Rome in 1701, where he applied himself particularly to mathematics and natural philosophy. In 1707 Galiani got hold of the Latin edition of Newton's *Opticks*, sent to Rome by the English envoy to Florence, Sir Henry Newton (no relation). The following year, Galiani performed a few of the experiments contained therein before the Accademia degli Antiquari Alessandrini, assisted by the director of the Academy, the astronomer Francesco Bianchini. The results of Galiani's experiments and written observations on Book I of the *Opticks* were circulated among friends. Galiani's embrace of Newtonian optics was soon broadened to include the mechanics. In 1708 he carefully read David Gregory's exposition of Newtonianism in his *Astronomiae physicae et geometricae elementa* ("Elements of Physical and Geometrical Astronomy") and then turned to the *Principia* itself. Galiani's manuscript "remarks" on several propositions of Book I indicate the difficulties he encountered in his private study, which, however, did not prevent him from accepting the premise that universal gravitation best accounted for the motion of celestial bodies. William Burnet's visit to Rome in 1709 helped deepen and ground Galiani's understanding of the *Principia* as well as put the monk in contact with Hermann and other northern mathematicians.[55] The opening of the Accademia Gualtieri in Rome in 1714 served as the occasion for Galiani's expounding before it the most forthright advocacy of Newton in Italy to date, in the form of the *Epistola de gravitate et cartesianis vorticibus* ("Letter on Gravity and Cartesian Vortices"), which demolished Cartesian physics while making a compelling case for Newtonian cosmology. Galiani sought to publish his treatise in the *Giornale de' Letterati d'Italia* – obviously to improve on Poleni's more circumscribed publication of some two years earlier – but the editors appear to have been reluctant to print such an inflammatory attack on Cartesianism. Not to be deterred, Galiani circulated the manuscript widely, and it served to convert quite a few savants to Newtonianism.[56]

Galiani has been described as the *éminence grise* of Newtonianism in Italy. "Newton was too much in his blood," his nephew once quipped as proof of his uncle's orthodoxy. But Galiani's centrality as proselytizer did not derive from any publication. A brush with censorship in 1711, and his subsequent realization of the delicacy of his scientific campaign, made Galiani eschew print altogether. From his cell at the Monastery of Sant'Eusebio in Rome, however, the Celestine monk spawned a network of epistolary exchanges that embraced much of the Italian peninsula and helped to turn Rome into an important center of scientific life. With Cardinal Gualtieri's encouragement, Galiani also made a concerted effort to coordinate communication between various Italian academies, notably the one established in Rimini by the future Cardinal Giannantonio Davia and managed by his

SAGGIO
DELLA
FILOSOFIA
DEL SIGNOR CAV.
ISACCO NEWTON
ESPOSTO CON CHIAREZZA
DAL SIGNOR
ENRICO PEMBERTON
Con una Differtazione dello steſſo su la misura della Forza de' Corpi in moto cavata dagli Atti Filosofici d'Inghilterra.
OPERA TRADOTTA DALL' INGLESE.
Aggiuntovi l'Estratto di altra dissertazione contraria su lo stesso Argomento.

IN VENEZIA, MDCCXXXIII.
Presso Francesco Storti In Merceria.
CON LICENZA DE' SUPERIORI, E PRIVILEGIO.

The Italian translation (Venice, 1733) of Henry Pemberton's *View of Isaac Newton's Philosophy* further contributed to the efforts of Galiani and his friends to disseminate Newtonian natural philosophy in Italy. – Burndy Library

This 1729 caricature from the Vatican Library shows Celestino Galiani (1681–1753), the Celestine monk who was Newton's greatest champion in early eighteenth-century Italy. – Biblioteca Apostolica Vaticana, Rome; courtesy of Paolo Casini

In this likeness of Newton from the Museo *La Specola*'s Figures Room, the Englishman is seen with his left hand resting on a copy of the *Principia*, while in his right hand he swings a small sphere linked to a larger one as if by a bridle, to signify action at a distance. – Courtesy of Museo *La Specola*, INAF-Astronomical Observatory of Padua, Italy

physician, Antonio Leprotti; the Bologna Institute of Sciences; and the group gathered around Guido Grandi in Pisa. Following his appointment as *capellano maggiore* in 1731, Galiani channeled his considerable energies to reconstitute Neapolitan science along the lines he had pursued in Rome – and with great success.

Activities such as Galiani's behind-the-scenes, sometimes even clandestine, efforts on behalf of Newtonianism were common among Italian savants, many of whom were his friends. Antonio Leprotti was one such person. Among his manuscripts one finds eight volumes devoted to "physics," attesting to Leprotti's close study of Newtonian science and his espousal of the chief tenets of the *Principia*. His letters, too, reveal a mind akin to Galiani's. The strength of the Newtonian philosophy, Leprotti pronounced in 1733, "has always seemed to lie entirely in Galileo's way of philosophizing." Unlike Descartes, Newton "did not produce a general system of physics, for he did not believe there was sufficient data on which to fabricate one. But he furnished some particular treatises, reducing a series of observations that he found had a common principle to a system.... One ought confess that this universe is too vast to wish to put it under a system manufactured by our minds." As for Locke, Leprotti confided to another correspondent, "there are some things regarding revelation and the confines of reason and faith which (to tell the truth) I would not know if they concur with our sacrosanct principles, but just since in this field I proposed the creed as the foundation, I do not worry about what Locke or the other philosophers say."[57] Leprotti's rise to prominence in Rome – he became the Pope's physician – obviously placed him in an important position to advance the cause of modern science in Italy as well as to counter the efforts of conservatives to arrest its progress.

Leprotti's assertion that Newtonian science lay "entirely in Galileo's way of philosophizing" attests to the relative ease with which many Italians embraced Newton as Galileo's most distinguished follower. The Tuscan martyr occupied the same position for Italians that Descartes occupied for the French. Yet, whereas Newton excoriated and dethroned Descartes, Galileo received lavish praise in the *Principia*, with Newton even ascribing to the Italian the discovery of his first two laws of motion. (Descartes, for his part, all but ignored Galileo in his published writings.) Newton, therefore, could be made to fit with relative ease into the Galilean experimental-mathematical tradition, so much so that the espousal of Newtonianism became a means to celebrate the national hero. Small wonder, then, that the direct lineage between the two became a familiar feature of contemporary Italian scientific and literary writings. Niccolò di Martino's historical preface to his *Elementa statica* ("Elements of Statics"), for example, depicted the *Principia* as the greatest fruit of Galilean science. Paolo Frisi concurred. "Galileo and Newton were bound to follow each other," he wrote in 1765; "both were free enough, enterprising, and active as to give a new form to science.... Both were equipped with all the necessary talents, the former to begin the scientific revolution, the latter to give it the form that they must keep stable."[58] Such sentiments would be echoed again and again.

Newton's celebrated eschewal of hypotheses and of metaphysics – as well as his incontrovertible credentials as an experimentalist –

also perfectly suited the political needs of a scientific community in a region where the religious and academic authorities vigilantly scrutinized its activities. Indeed, a desire to keep separate the domains of science and religion – an ideal championed by Galileo – prompted many savants to pursue a line of research that steered clear of investigation into the causes of things, and that focused instead on the phenomena of nature, just as Newton had advocated. Within a few decades, such a research course would assure the spectacular institutionalization of science in Bologna and the particularly high technical level of research carried out there.[59] But savants in other centers embarked on a similar course.

By the third and fourth decades of the eighteenth century, growing numbers of teachers came to incorporate Newton into their teaching of natural philosophy, often stressing the phenomenological modesty of Newtonian science. In the preface to the 1745 Italian translation of Musschenbroek's *Elementa physicae*, for example, Antonio Genovesi pronounced the merit of following Newton, who "wanted to philosophize, not with vague hypotheses and conjectures, but with conclusive experiments and solid reasoning, confirmed by experiments." For that reason, Genovesi continued, physics throughout Europe moved away from Cartesian "romance" to "a firmer and more solid" base. On such grounds, many textbook writers, particularly priests, even managed to fit Newton into an overarching Aristotelian framework. The Somascan father Giovanni Francesco Crivelli is a case in point. In 1731 – two years before his election as a Fellow of the Royal Society – Crivelli published his influential *Elementi di fisica* ("Elements of Physics"), which included a learned and detailed account of both the *Principia* and the *Opticks*. Crivelli refrained from taking sides except when rejecting notions that were flagrantly at odds with the scholastic/doctrinal framework, for example, the notion of a void. Even so, his balanced presentation presented old and new theories, side by side, inviting readers to draw their own conclusions.[60]

Circumspection and decorum thus eased the way for the teaching and publication of Newtonian ideas in Italy even among Jesuits. Roger Boscovich mastered the technique. His first public presentation as professor of mathematics at the Collegio Romano in 1740 focused on the aberration of light. To assuage any concerns of his auditors and superiors regarding his choice of cosmology, Boscovich declared that he would set aside Scripture for the purpose of his presentation and avail himself of heliocentrism because of its simplicity (he undoubtedly recognized that the newly discovered phenomenon could hardly make sense otherwise). In this and other works, Boscovich, like Crivelli, strove to demonstrate that Newtonianism could be made to fit into an Aristotelian framework. As a Newtonian, Boscovich could also dodge identification as a Copernican. Such maneuvering allowed him to declare that "with respect due to my state as a Christian, as a catholic and as a religious, I managed to separate the philosophical question from Christian polemics." Even when boasting of his discovery of "a new kind of Universal Natural Philosophy, one that differs widely from any that are generally accepted and practised at the present time" – as he did in the dedication of his celebrated *Theoria philosophiae naturalis* ("A Theory of Natural Philosophy") – Boscovich maintained

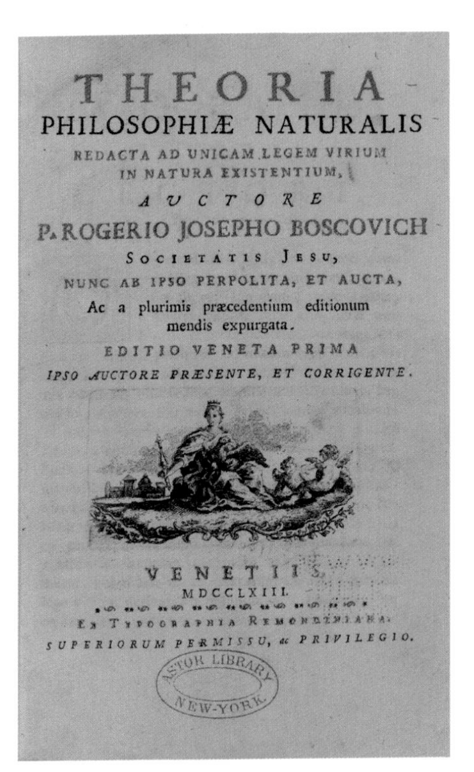
THEORIA
PHILOSOPHIÆ NATURALIS
REDACTA AD UNICAM LEGEM VIRIUM
IN NATURA EXISTENTIUM,
AUCTORE
P. ROGERIO JOSEPHO BOSCOVICH
SOCIETATIS JESU,
NUNC AB IPSO PERPOLITA, ET AUCTA,
Ac a plurimis præcedentium editionum
mendis expurgata.
EDITIO VENETA PRIMA
IPSO AUCTORE PRÆSENTE, ET CORRIGENTE.
VENETIIS,
MDCCLXIII.
SUPERIORUM PERMISSU, ac PRIVILEGIO.

Despite his self-proclaimed position halfway between Newton and Leibniz, Roger Joseph Boscovich modeled his *Theoria philosophiae naturalis* on the *Principia*, further announcing his fulfillment of Newton's dream to unify all forces of nature. This is the second edition (Venice, 1763). – NYPL–SIBL

Celestino Cominale's *Anti-Newtonianismi* (Naples, 1754–56), from which this plate comes, was a passionate, though not particularly well-informed, attack on Newton's optics and universal gravitation. – NYPL–SIBL

that nothing in his synthesis of the theories of Newton and Leibniz was not "suitable in a high degree, for the consideration of a Christian priest." On the contrary, he considered his work to be "in complete harmony with it."[61]

This sketchy account of the early diffusion of Newtonianism into Italy does little justice to the full range of the dissemination, the details of which still await research. Nor should it be assumed that the diffusion was free of opposition, religious or scientific. Far from it. As occurred elsewhere in Europe, the challenges presented by the recondite nature of the *Principia* could cause even professionals to despair. The Bolognese astronomer Eustachio Manfredi, for one, commented in 1716 that he eagerly awaited the completion of Giuseppe Verzaglia's commentary on the *Principia*, for it promised to render its content clearer and more methodical to those, like himself, for whom Newton appeared to be "speaking Arabic." A decade later, the physician Giovanni Bianchi admitted that his inability to understand universal gravitation made him prefer the concept of Cartesian vortices, which, to his mind, "although defective in some points, was the best suited to explain natural phenomena, unlike the most obscure and defective hypothesis of Newton."[62]

Cartesianism, however, was less entrenched in Italy than it had been in France, and its viable opposition to Newtonianism was short-lived. More important was opposition based on the perceived connection between Newtonianism and the materialist-oriented philosophy of John Locke. Antonio Conti attempted to dispel precisely such a danger while establishing the congruence of the two Englishmen's ideas: "Newton has said nothing that is not original, and he has expounded nothing that he has not rigorously demonstrated.... Without doubt all that is sounder and more original in Locke's *Essay on Human Understanding* comes from him. To Sir Newton must be ascribed, whether in England

This allegorical frontispiece to volume 3 of Jacopo Riccati's collected works shows Urania holding up the mirror reflecting Truth, her right foot resting on a sphere, her left defeating ignorance. - NYPL-SIBL

or in France, that one is willing to admit in philosophy only experience, observation, and mathematical reasoning." A decade and a half later, Doria found cause for alarm in this very genealogy. "The modern philosophers have been searching for another new science. They first clung to the doctrine of Mr. Newton, but because that great mathematician and philosopher did not concern himself much with metaphysics, many of the moderns then stopped at the philosophy of Mr. Locke.... And this is now the sect that is taught by many teachers and students at Rome, at Naples, and in many other parts of Italy. As a result it has a good number of followers."[63]

In retrospect, then, it becomes increasingly clear that critics were concerned not so much with the content of the *Principia* or the *Opticks* as with the extrapolation of Newtonian ideas to sanction ideological ends. For this reason, criticism along religious lines hardly diminished the admiration of Newton's genius, or hindered the dissemination of his ideas. Jacopo Riccati is a case in point. His advocacy of Newtonian mechanics was unwavering, evident even in works in which he opposed Newton on metaphysical grounds. His *Saggio intorno il sistema dell'universo* ("Essay About the System of the Universe"), for example, endorsed the physics

Francesco Algarotti (1712–1764) was only sixteen years old when, under the guidance of Francesco Maria Zanotti, he replicated Newton's prismatic experiments at the Bologna Institute of Sciences. – NYPL–Music Division

of the *Principia* while simultaneously rejecting its perceived natural theology. As Riccati stated at one point, "without prejudice against his sublime discoveries, it seems to me that Sir Newton could avoid the daring expression that space is God's sensorium." This attempt to rid Newton's physics of its seeming fusion of divinity and infinite space, substituting instead a more palatable metaphysics, obviously somewhat diminished the coherency and the consistency of Riccati's treatise. Nevertheless, his approach illustrates the manner in which it became possible to graft different metaphysical positions onto the *Principia* and the *Opticks* in order to facilitate their accommodation into different ideological contexts.[64]

Riccati's partial support of Giovanni Rizzetti's criticism of the *Opticks*, in contrast, was motivated primarily by friendship. Both men were noblemen from Castlefranco, born a year apart, whose education, class, family, and scientific interests drew them closely together. Rizzetti, however, lacked not only Riccati's mathematical acumen but also his commitment to a professional life in science. Rizzetti was, and remained, an amateur, whose vanity and pride prevented him from ever acknowledging his shortcomings or errors.

Rizzetti's critique of Newton was triggered by the failure of Francesco Maria Zanotti – professor of philosophy at Bologna and secretary of the Institute of Sciences there – to replicate in 1720 Newton's prismatic experiments, principally because the prisms at his disposal were defective. Emboldened by this failure, Rizzetti composed a paper that not only challenged Newton's theory of colors, but controverted the verity of certain of his experiments. Newton took the trouble to draft a reply (which he never sent) as well as to instruct the Royal Society's Curator of Experiments, John Theophilus Desaguliers, to perform, once

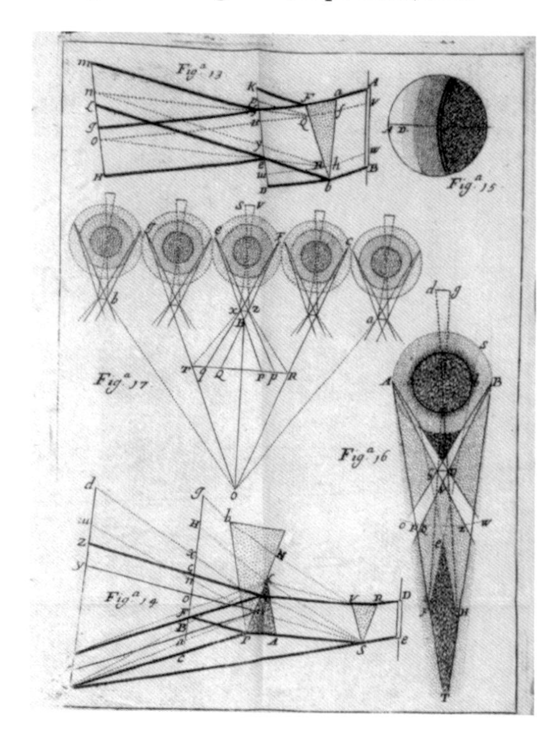

Giovanni Rizzetti never relented in his anti-Newtonian campaign. His 1727 *De luminis affectionibus specimen physico mathematicum*, which includes this plate, was followed in 1741 by a broader (and more violent) critique, in *Saggio dell'anti-newtonianismo sopra le leggi del moto e del colori*. – Burndy Library

In "Newton Discovers the Refraction of Light," Filippo Pelagio Palagi depicts Newton as if inspired in the presence of his half-sister and her child playing with soap bubbles by the window. – By permission of Civici Musei d'Arte e Storia di Brescia

again, the contested experiments before the Society. Their success failed to satisfy Rizzetti, who proceeded to publish his paper in the *Acta Eruditorum*, but his claims did not fare much better in Germany. Georg Friedrich Richter, the Leipzig professor of mathematics, immediately published a rebuttal in the same periodical. Rizzetti remained recalcitrant, publishing in 1727 a book that attacked Newton even more virulently. His extreme ideas and reckless charges effectively ousted him from the European scientific community.

At home, even Riccati, who was willing to endorse a methodological claim made by Rizzetti (and by Hooke and Huygens earlier) regarding the inherent dogmatism of Newton's insistence that his "crucial experiment" could have only one interpretation, refused to endorse Rizzetti's attack on Newton's experiments, or his alternative modification theory. In any event, by 1728 the young Francesco Algarotti had successfully replicated Newton's experiments before the Bologna Institute of Sciences, thus marginalizing Rizzetti even among local savants.[65]

4

A literary giant aloof from academic disputes, and an amateur in matters of science, François-Marie Arouet de Voltaire (1694–1778) demonstrated the accessibility of Newton's ideas to nonspecialists. – NYPL–Print Collection

THE VOLTAIRE EFFECT

The steady diffusion of Newtonian ideas into Europe before Newton's death is, then, undeniable. But in France, as we have seen, it was confined primarily to the domain of mathematicians and natural philosophers, as well as to a small circle of educated men who had acquired a taste for such studies. Wider dissemination required the arrival on the scene of the unique genius of Voltaire. What made Voltaire so effective an agent – apart from an unparalleled ability to seduce an audience by a masterful combination of shock and wit – was his disinterested status. Neither mathematician nor physicist, but a literary giant aloof from the academic disputes that had embroiled the previous generation, Voltaire's voice carried force. In other words, Voltaire's stature as an amateur in matters of science was the source of his contemporary appeal, demonstrating for the first time the accessibility of Newton's ideas to nonspecialists.

In truth, Voltaire's progress toward Newtonianism was gradual. In his youth, he spent seven years under the tutelage of the Jesuit fathers at the Collège Louis-le-Grand, an experience he often denigrated in later years. Dining with the Catholic poet Alexander Pope and his mother, Voltaire stunned his hosts by disclosing how "those d—d Jesuits, when I was a boy, b[u]g-g[a]red me to such a degree that I shall never get over it as long as I live." Less shocking was the course of studies he recounted: "not a word of mathematics, not a word of sound philosophy. I learned Latin and nonsense."[1] In actual fact, the Jesuit college offered sound mathematical and philosophical training for those so inclined; Voltaire, alas, was not among them. His passions were poetry and the theater, in which he indulged fully.

Nor did Voltaire evince much interest in the sciences for nearly a decade after his departure from the Jesuit College. Only in the early 1720s were Voltaire's philosophical interests piqued, in no small part owing to his acquaintance with the English philosopher and politicianViscount Bolingbroke, then in exile

In his *Astronomie physique* (Paris, 1740), from which this vignette is taken, Etienne-Simon de Gamaches declared his determination to defend Descartes' principles of celestial mechanics, undeterred by the authority of so great a genius as Newton. - NYPL-SIBL

in France. In April 1724, Voltaire received a long letter from Bolingbroke that would exert a significant influence on the young poet. Sounding almost like a schoolmaster, Bolingbroke exhorted Voltaire to cultivate his mind and to join reason to that admirable imagination that nature had bestowed upon him. Specifically, he urged Voltaire to wade through John Locke's *Essay on Human Understanding*, which would make abundantly clear to him that Descartes in his physics, and Malebranche in his metaphysics, had shown themselves to be poets rather than true philosophers. The imprudence of Descartes in sidestepping the boundaries of observation and geometry – believing them too narrow – had been subsequently exposed by the powerful minds of Huygens and Newton, who demonstrated the falsehood of the Cartesian laws of motion, and proved "that those famous vortices are chimeras." Malebranche, in Bolingbroke's equally critical opinion, had merely produced books filled with "the most beautiful nonsense in the world."[2]

Bolingbroke's letter has been rightly regarded as prefiguring "every important ideological position" staked out by Voltaire in the *Lettres philosophiques* ("Philosophical Letters" or "Letters on England"), published a decade later, in 1734.[3] At the time, Voltaire was still reveling in the extraordinary success of his early poetic and dramatic works. In early 1726, however, an unfortunate squabble over honor with the Chevalier de Rohan – which earned Voltaire the double ignominy of being beaten by the Chevalier's servants and being incarcerated in the Bastille – ended with his exile to England, where he remained for two and a half years. This unexpected sojourn abroad had momentous consequences both for the course of Newtonianism and the career of Voltaire. As one biographer put it: "Voltaire left France a poet, he returned to it a sage." Whereas before his flight "he had been actively productive in the sphere of the imaginative faculties," when he returned, "while his poetic power had ripened, he had tasted the fruit of the tree of scientific reason."[4]

The very choice of England attests to Voltaire's partiality for a country he would idealize as the seat of freedom, tolerance, and the philosophical spirit. In England, he wrote a friend at the outset of his visit, "one thinks

The orrery, a mechanical model of the solar system, was a popular device to exhibit the relative size and motion of the heavenly bodies. This engraving is from Part IV of *The General Magazine of Arts and Sciences* (London, 1755–64). – NYPL–General Research Division

freely and nobly," without being restrained "by any servile fear." Could he follow his inclination, he swore, he would remain in England if only "to learn how to think." So, too, he regarded English freedom as the source of the nation's greatness. "Fond of their liberty," the English are "a nation of philosophers," he confided to his friend several months later. French "folly" might well be "pleasanter than English madness, but by god English wisdom and English Honesty" are superior. By April 1728 he was boasting to another correspondent: "I think and write like a free Englishman."[5]

Voltaire never met Newton. Indeed, he got no closer than the magnificent funeral England accorded its distinguished son. The indelible impression the spectacle made on him was conveyed in the *Lettres philosophiques*. Newton "lived honoured by his compatriots and was buried like a king who had done well by his subjects," he wrote; "the highest in the land vied with each other for the honour of being pall-bearers." And how he envied a country that so revered and rewarded intellect: "Go into Westminster Abbey. It is not the tombs of kings that one admires, but the monuments erected by a grateful nation to the greatest men who have contributed to her glory."[6] To compensate for his failure to meet the great man himself, Voltaire befriended members of Newton's inner circle, key among them Catherine Barton, Newton's niece, and her husband, John Conduitt, as well as Sir Hans Sloane, president of the Royal Society. Crucial for his intellectual ripening was the friendship Voltaire struck up with Samuel Clarke, the renowned divine and philosopher who initiated him into the metaphysical and religious foundations of Newtonian philosophy. Unquestionably, Voltaire's deism was influenced by his discussions with Clarke, as were his future conceptions of time and space, free will, and proofs for the existence of God.[7] Voltaire also made the acquaintance of Henry Pemberton, editor of the third edition of the *Principia* and the author of *A View of Sir Isaac Newton's Philosophy*. Indeed, their familiarity was such that by May 1727, Voltaire had read Pemberton's *View* in manuscript and warmly recommended its translation into French – almost a full year before publication.[8]

In his *Discours sur les differentes figures des astres* (Paris, 1732), Pierre-Louis Moreau de Maupertuis controverted the prevailing opinion in France that universal gravitation was an "absurd principle," insisting that it was metaphysically viable and mathematically superior to vortices as an explanation of celestial mechanics. Shown here is the frontispiece to the 1742 edition. – Burndy Library

The manuscript notebooks that Voltaire kept during these years attest to his gathering of material for a projected account of his sojourn to England as well as to his gradual – if not as yet complete – conversion to Newtonianism. On at least one occasion he sounded a note of skepticism: "Even if attraction were true," he jotted down, "it would not provide the least advantage or the least assistance in mechanics. Yet Newton spent his life in pursuing this discovery and thousands of people in understanding it." In a more mischievous vein, he appropriated Pascal's wager argument – which provided pragmatic reasoning for belief in God – to rationalize his own budding Newtonianism. By taking "Newton's side," Voltaire noted, he had hedged his bets like the man who bequeathed money for masses to be said for him, with the proviso that if these failed to produce the desired effect, the money should revert to other charitable purposes. More tellingly, Voltaire contrasted Descartes and Newton, and found the former wanting. True, Descartes had made certain discoveries: he was the first to understand the nature of refraction and the first to perfect telescopes – but it was left to Newton to discover why telescopes could not be further perfected. In another series of notes, Voltaire enumerated Descartes' "errors" in the life sciences en route to summarily dismissing as nonsense (*billevesées*) the philosopher's notions regarding innate ideas, automata, emission of light, and his *Principia philosophiae* more generally. Newton, in contrast, emerged as an unequalled genius: "Before Kepler all men were blind; Kepler was one-eyed, and Newton had two eyes."[9]

Early public testimony on the direction Voltaire's mind (and heart) were taking can be glimpsed in the preface to the 1727 *Essay on Epic Poetry*, in which Voltaire stressed the unique perspective informing his projected account of his stay in England: "it strikes my Eyes as it is the land which hath produced a Newton, a Lock, a Tillotson, a Milton, a Boyle," men whose glory would "not be confined to the Bounds of this Island." Discussing Milton's genius in the body of the work, Voltaire inserted – for the first time – the famous anecdote regarding the apple: just as Milton derived his inspiration for *Paradise Lost* from a silly play he had seen in Italy, so Newton, "walking in his Gardens had the first Thought of his System of Gravitation, upon seeing an Apple falling from a Tree." The following year, in his dedication to Queen Caroline of the *editio princeps* of *La Henriade*, Voltaire provided another signal of his sentiments: "Our Descartes, who was the greatest Philosopher in Europe, before Sir Isaac Newton appeared," dedicated his *Principia* to Princess Elizabeth, not because she was a Princess but because "she understood him the best." On similar grounds, Voltaire dedicated his book to Queen Caroline, who was not only the "Protectress of all Arts and Sciences," but "the best Judge of them." Two years later, back in France, Voltaire published a new edition of his epic poem. Prudently, he removed the dedication to the English queen; yet, significantly, the cosmology of vortices that had informed canto seven of the earlier version now gave way to universal gravitation:

> These stars bound by the law that
> propels them,
> Attract and endlessly avoid each other
> on their path,
> And, serving one another of both rule
> and support,
> Make use of the illuminations they receive
> from him.[10]

It is noteworthy that at this stage Voltaire still believed it necessary to attach a flippant note of disclaimer: "Whether or not one accepts Mr. Newton's attraction, it is nevertheless certain that the heavenly spheres seem to attract or to repel one another."

Voltaire returned to France in early 1729 a philosophical poet, his head abuzz with an assortment of new interests that for the next decade threatened to overtake his poetic vocation. By this date he had also probably drafted his letters concerning English religion, government, and literature. Only in the course of 1732, however, did his letters on Newton (and Locke) acquire their final shape, following an intense exchange with Pierre-Louis Moreau de Maupertuis, who had just published his pro-Newtonian *Discours sur les differentes figures des astres* ("Discourse on the Various Shapes of the Celestial Bodies"). Maupertuis assuaged Voltaire's few remaining doubts regarding

SUR LES ANGLOIS. 105

QUATORZIE'ME

LETTRE

SUR

DES CARTES

ET

NEWTON.

UN François qui arrive à Londres, trouve les chofes bien changées en Philofophie comme dans tout le refte. Il a laiffé le monde plein, il le trouve vuide. A Paris on voit l'Univers compofé de Tourbillons, de Matiere fubtile; à Londres on ne voit rien de cela. Chez vous c'eft la preffion de la Lune qui caufe le flux de la mer; chez les Anglois c'eft

The fourteenth letter in Voltaire's *Lettres philosophiques* offered a concise and witty contrast between the scientific mentalities of Descartes and Newton: "For your Cartesians everything is moved by an impulsion you don't really understand, for Mr. Newton it is by gravitation, the cause of which is hardly better known. In Paris you see the earth shaped like a melon, in London it is flattened on two sides." Shown here is the first page of the letter from an edition published in 1734 as *Lettres ecrites de Londres sur les Anglois.* – NYPL–Rare Books Division

universal gravitation as well as read through the Newtonian letters to ascertain their scientific accuracy. By 1733, the manuscript of the *Lettres philosophiques* was in the hands of printers in both London and Paris, but publication was delayed when royal permission was not forthcoming. Ultimately, two different English editions appeared later that year without Voltaire's consent, followed in April 1734 by a surreptitious Paris edition. The book enjoyed an immediate *succès de scandale*: a warrant for Voltaire's arrest was issued the following month, precipitating the author's hasty flight to Mme du Châtelet's chateau at Cirey. In the author's absence, the authorities took their revenge on the Parisian publisher – throwing him into the Bastille – as well as on the book, ordering it publicly burned by the common hangman on June 10.

Obviously, the authorities were enraged by what they perceived to be willful disregard of royal displeasure. Beyond procedural issues, however, the book struck a raw nerve. Voltaire had conjured up a hegemonic model of civilization that, at least by implication, rendered contemporary French society deficient. The emphasis on the healthy state of religious pluralism in England, the implications for religion of Locke's ideas regarding "thinking matter," and the explosive final chapter on Pascal, all turned the author into a provocateur. The Newtonian chapters were equally incendiary, framed in a manner that demeaned Descartes, if not ridiculed him outright. "Our Descartes," Voltaire opined in the letter on Locke, "born to uncover the errors of antiquity but to substitute his own, and spurred on by that systematizing mind which blinds the greatest of men, imagined he had demonstrated that the soul was the same thing as thought, just as matter, for him, is the same thing as space." The letter on Descartes and Newton, which opens with the celebrated contrast between the scientific mentalities of the two men, deserves to be cited more fully:

A Frenchman arriving in London finds things very different in natural science as in everything else. He has left the world full, he finds it empty. In Paris they see the universe as composed of vortices of subtle matter, in London they see nothing of the kind. For us it is the pressure of the moon that causes the tides of the sea; for the English it is the sea that gravitates towards the moon, so that when you think that the moon should give us a high tide, these gentlemen think you should have a low one.... Furthermore, you will note that the sun, which in France doesn't come into the picture at all, here plays its fair share. For your Cartesians everything is moved by an impulsion you don't really understand, for Mr Newton it is by gravitation, the cause of which is hardly better known. In Paris you see the earth shaped like a melon, in London it is flattened on two sides. For the Cartesians light exists in the air, for a Newtonian it comes from the sun in six and a half minutes. Your chemistry performs all its operations with acids, alkalis and subtle matter; gravitation dominates even English chemistry.

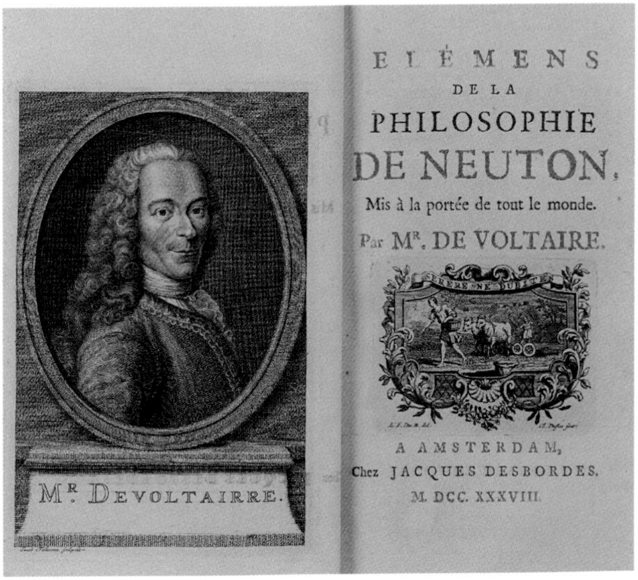

The contrast between the life and thought of the two savants leaves little doubt of Voltaire's preference. True, Descartes is credited with important contributions to mathematics; all his other works, however, "are full of errors." Having abandoned mathematics as its guide, Descartes' "philosophy was nothing more than an ingenious novel." Equally egregious, he "created a world and made man to his own specification, and it is said, rightly, that Descartes' man is only Descartes' man and far removed from true man." Having thus cut Descartes down to size, Voltaire devoted two letters to an explication of Newtonian mechanics and optics.[11]

The *Lettres philosophiques*, Voltaire observed with both irony and foresight in 1733, was penned in a "spirit of liberty which, perhaps, will bring upon me persecutions in France." The difference between the English and the French was such, he went on to remark, that whereas readers in London regarded the letters as "philosophical," in Paris they were regarded as "impious," sight unseen.[12] The reception of the volume – though briefly proving Voltaire correct insofar as persecution was concerned – soon dispelled his pessimism regarding the response of the French reading public. If he worried in 1733 whether the book would sell, in light of the prejudices of the French and the paltry 200 copies sold of Maupertuis' *Astres*, he soon discovered that his endeavor "to brighten the dryness of these matters and make them palatable to the taste of the nation" paid rich dividends: within five years, the book had sold some 20,000 copies.[13] Wide readership, of course, did not necessarily translate into wholesale conversion of the French reading public to Newtonianism. The controversy – and curiosity – the book elicited assured instead a more auspicious atmosphere that would greet Voltaire's next philosophical publication, the *Elémens de la philosophie de Neuton* ("Elements of Newton's Philosophy").

The *Elémens* was probably not on Voltaire's mind when he put the finishing touches on the *Lettres philosophiques*. His final letter to Maupertuis in December 1732 concluded by articulating his resolve to renounce this course of studies: "Adieu; I love and admire you, but I'm obliged to abandon this philosophy: this is a calling [*métier*] that demands great health and leisure, and I have neither the one nor the other."[14] Several factors, however, conspired to keep Voltaire's attention focused on science and philosophy. First, charges of impiety, explicit or insinuated, hurled against him made Voltaire anxious, and he was determined to publish an apologia in the form of a treatise on metaphysics. His

It took Voltaire barely a year to "perfect his understanding of Newton's mechanics and optics, and to write the *Elémens de la philosophie de Neuton*, one of the most successful Enlightenment "popularizations of Newton's ideas. This edition was published in Amsterdam in 1738. – NYPL–Rare Books Division

In a letter, Voltaire acknowledged the important contribution that his lover and collaborator, Mme du Châtelet, made to the composition of his *Elémens de la philosophie de Neuton*: "Minerva dictated and I wrote." The allegorical frontispiece to the book conveys precisely that message, as du Châtelet reflects the rays of light (truth) emanating from the heavens behind Newton onto the inspired Voltaire, who is busy at work below.
– NYPL–Rare Books Division

enforced domicile at Mme du Châtelet's chateau at Cirey provided him with the requisite leisure to carry out such a project. Indeed, his liaison with Mme du Châtelet had even greater consequences for his philosophical work. Her growing immersion in scientific studies, aided and abetted by Voltaire and others, could not but draw him in that direction as well. So, too, the arrival at Cirey in October 1735 of the young Italian count Francesco Algarotti, who charmed his hosts with his personality as well as with his manuscript *Newtonianismo per le dame* ("Newtonianism for Ladies"), kindled Voltaire's desire to emulate (and surpass) Algarotti in such an undertaking. Predictably, Voltaire produced a more profound work of popularization, but the structure of the *Elémens* bears witness to the influence of Algarotti.

In truth, Voltaire was predisposed to the idea of embarking on a popularization of Newton even before he met Algarotti. When he returned to Paris in April 1735 for the first time since fleeing the capital nearly a year earlier, he noted that not only had Newton's philosophy gained favor "among the true philosophers" since the publication of the *Lettres philosophiques*, but the general taste in the capital had changed:

> ... verses are no longer fashionable in Paris. Everybody tries to be a geometer or a physicist; people dabble with reason. Sentiment, the imagination, and grace are banished. Were a man who had lived under Louis XIV to return to this world, he would not have recognized the French and think the Germans had conquered the country. The belles-lettres perish from sight.[15]

Voltaire took some credit for effecting this change in sensibilities – though, as the above quotation suggests, he was unwilling to condone the tyranny of philosophy over literature. Certainly, his awareness of his uncanny ability to sway the minds and tastes of contemporaries played into his decision to embark on the *Elémens*.

Incredible as it may seem, the entire process of perfecting Voltaire's grasp of Newtonian science and of writing the book took barely a year (July 1736–Summer 1737). Friends offered assistance by way of furnishing books and advice. Voltaire had a laboratory fitted at Cirey and carried out numerous experiments there. Above all, he relied on the counsel of Mme du Châtelet – so much so that he would later represent the *Elémens* as virtually a joint venture. As he intimated to the future Frederick II on January 15, 1737: "I've sketched easily enough the principles of Newton's philosophy and Mme du Châtelet had a part in the work. Minerva dictated and I wrote."[16]

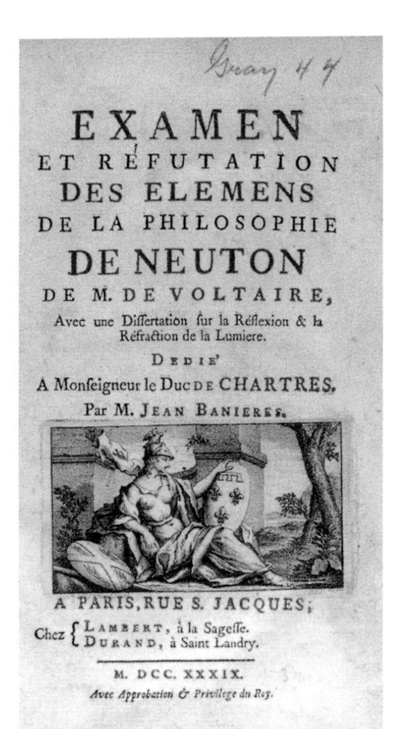

EXAMEN
ET RÉFUTATION
DES ELEMENS
DE LA PHILOSOPHIE
DE NEUTON
DE M. DE VOLTAIRE,
Avec une Diſſertation ſur la Réflexion & la Réfraction de la Lumiere.
DEDIE'
A Monſeigneur le Duc DE CHARTRES,
Par M. JEAN BANIERES.

A PARIS, RUE S. JACQUES;
Chez { LAMBERT, à la Sageſſe. DURAND, à Saint Landry.
M. DCC. XXXIX.
Avec Approbation & Privilege du Roy.

Jean Banières' *Examen et réfutation des elemens de la philosophie de Neuton de M. de Voltaire* (Paris, 1739) offered a caustic critique of the *Elémens* of Voltaire. – Burndy Library

Voltaire had precisely such sentiments in mind when he designed an allegorical frontispiece for the book. There, attired in a Roman toga, the poet's laurels resting on his head, books and mathematical instruments all around him, Voltaire sits at his desk composing the *Elémens*. Directly above him is Newton, regally seated on a throne of clouds, his right hand positioning a compass on a celestial globe, the index finger of his left hand pointing at the globe. All the while his gaze fixes on Mme du Châtelet – levitating halfway between Newton and Voltaire, thanks to some helpful putti – who returns the gaze. The dynamics of the scene seem to suggest Newton the master impressing an important lesson on his admiring Emilie. More telling still, the oval mirror in du Châtelet's grip collects the rays of light (truth) emanating, as it were, from the heavens behind Newton – and reflects them onto the inspired Voltaire, busy at work below.[17]

The *Elémens* proved an immense, instant success, for the most part free of the scandal – if not controversy – that surrounded the *Lettres philosophiques*. Even those men of letters not predisposed to Voltaire recognized the magnitude of his achievement, and said as much publicly. The *Bibliothèque Françoise*, for example, devoted thirty-one pages to the book; "one can sense the reviewer's admiration being wrung from him against his own inclinations." The reviewer for the Jesuit *Journal de Trévoux* – almost certainly that implacable opponent of Newtonian ideas Louis-Bertrand Castel – proved surprisingly flattering: "Nothing proves better the cutting efficacy of the printed word, and the superiority of a man who knows how to handle it. Newton, the great Newton, was, it is said, buried in the abyss for twenty-seven years, in the shop of the first publisher who dared to print him." A thousand savants descended into the abyss in an endeavor to make Newton comprehensible, the reviewer continued, but only the learned became enlightened by their efforts. "Voltaire finally appeared, and at once Newton is understood or is in the process of being understood; all Paris resounds with Newton, all Paris stammers Newton, all Paris studies and learns Newton."[18]

The book had its critics, of course, mostly die-hard academic Cartesians, such as Jean Banières, Noël Regnault, and Castel. As for Voltaire, he seemed to enjoy the task of refuting cavilers, at times appearing almost eagerly to embrace the role of martyr as well as iconoclast. When in October 1739 he sent a copy of his "Réponse à toutes les objections principales qu'on a faites en France contre la philosophie de Neuton" ("Response to All the Principal Objections That Were Made in France Against Newton's Philosophy") to Martin Folkes – soon to be elected president of the Royal Society – Voltaire proclaimed in English that he was "obliged to write against our antineutonian cavillers," likening himself to "a man blind of one eye expostulating with stark blind people who deny there is such Thing as a sun." Four years later, upon his election as a fellow of the Royal Society, the grateful Voltaire reprised his leading role: "I made some steps ... in the temple of philosophy towards the altar of Newton. I was even so bold as to introduce into France some of his discoveries; but I was not only a confessor to his faith, I became a martir." As late as 1768, long after his interest in science had waned, he was still proud of the role he had been called upon to play: "My fate has decreed that I should be the first of my countrymen to be permitted to explain the discoveries of the great Newton. I have been the apostle and the martyr of the English."[19]

Pierre Jose Perrot's *éloge* of Newton was published in Eugene Mac Swiny's *Tombeaux des princes* (Paris, 1741). – NYPL–Art & Architecture Collection

The religious metaphors deployed by Voltaire are seminal to his conception of Newtonian science and to his awareness of his evangelical mission. Newtonianism became something of a secular religion for Voltaire, as witnessed by the traditional religious terminology he harnessed to his cause. In 1732, for example, he confided to Newton's "apostle" Maupertuis that Maupertuis' first letter resolving his doubts concerning universal gravitation had "baptized" him; the second had "confirmed" him. "My faith depends on yours," he avowed; "I am your proselyte and take my profession of faith from your hands." Four years later, at work on the *Elémens*, he informed a correspondent that Newton "is here the God to whom I sacrifice, though I have some chapels to other lesser deities."[20] With the publication of the *Elémens*, Voltaire came to regard himself, too, as an apostle. Writing to a friend in August 1738, Voltaire invoked Matthew 19:14 – "But Jesus said, Suffer little children, and forbid them not, to come unto me: for of such is the kingdom of heaven" – to claim that the "kingdom of heaven" was Newton, and that the French were the little children. Understandably, such posturing led him to prophesy later in the same letter that "the Regnaults and the Castels [would] not be able to prevent for long the triumph of reason."[21]

Voltaire came very close to idolizing Newton. "Divine" creeps in as an attribute of the Englishman, and Voltaire's veneration approaches faith. "Newton is the greatest man who has ever lived," he told a former teacher, abbé d'Olivet, in 1736; compared to Newton, the giants of antiquity were but children who play games.[22] This glorification of the man and his science had significant consequences for Voltaire's propagandist efforts. The two Newtonian tablets, the *Principia* and the *Opticks*, became tantamount to gospel; the languages in which they were written, those of mathematics and experiment, were inspired and irrefutable. These evangelical overtones informed his attempted conversion of an acquaintance in March 1739: "The faith that I would dare ask you is for certain indispensable calculations, for certain demonstrated propositions; after which we shall be of the same religion, and I will have honor to doubt with you seven or eight thousand propositions, so long as you granted me only a dozen truths founded on the experiment." In retrospect, at least, Voltaire's manner of preaching evokes the self-assurance of d'Alembert as he urged on those struggling with the calculus: "Go forth, and faith will come to you."[23]

The accessibility of Voltaire's *Elémens* to laymen is not easy to ascertain. The title page of the 1737 Dutch edition proclaimed that the book now brought Newton's philosophy "within everybody's reach." Voltaire pretended displeasure with such egalitarianism, fearing it smacked of false advertising or, worse still, seemed aimed at children. Voltaire himself had criticized Algarotti's *Newtonianismo* – which he received shortly after the *Elémens* was published – for being "charming but shallow, gathering all of Newton's blooms and leaving Voltaire the thorns." Ten pages of the *Elémens*, he was certain, contained "more truth" than Algarotti's entire book. Nevertheless, while at work on his own book, Voltaire "insisted that he wanted to be understood by the common people and to that end he had sought to make his exposition of Newtonianism 'as clear as a fable of La Fontaine.'"[24] Some commentators nevertheless found the *Elémens* specialized. The Marquis d'Argens opined that at least a modicum of mathematics was necessary for comprehending certain demonstrations of the *Elémens* and thus estimated that no more than three thousand Frenchmen were capable of benefiting from it.[25]

Be that as it may, Voltaire's impact on the reception of Newtonianism by professional and lay audiences alike is beyond dispute. Writing in the 1770s of Maupertuis' introduction of Newtonianism to the Paris Académie des Sciences, one observer noted that by his *Lettres philosophiques* and by the *Elémens* "Voltaire contributed not a little to this revolution." Voltaire, too, believed in his success, although he quipped in 1759 that the establishment in France of Newton's sublime truth had required first the passing of the entire generation that had grown old in Descartes' errors.[26] Condorcet concurred. As late as 1774 he was certain that the *Elémens* was "still the only book in which men who have not cultivated the sciences can acquire simple and precise notions concerning the system of the world, and about the theory of light." As Condorcet explained to a correspondent, the book was published at a time when "Jean [Johann] Bernoulli, the greatest mathematician in Europe, still opposed Newtonianism; more

French philosopher, mathematician, and politician Marie Jean-Antoine-Nicolas Caritat, marquis de Condorcet (1743–1794), considered Voltaire's *Elémens* to be "still [in 1774] the only book in which men who have not cultivated the sciences can acquire simple and precise notions concerning the system of the world, and about the theory of light." – Smithsonian Institution Libraries, Washington, D.C.

than half of the Académie des sciences was Cartesian; even Fontenelle ... who was one of the few able to understand it, remained obstinately attached to his original opinion. If to all this is added the fact that the French schoolbook in which the theories of Newton were expounded did not appear until ten years" after the *Elémens*, "one cannot but agree that there was much merit in publishing in 1738" what Voltaire called "with so much modesty his little catechism of gravitation."[27]

Descartes vs. Newton: The Ground Shifts

Only difficulties with the censors had prevented the *Elémens* from appearing on the centenary of the publication of Descartes' *Discours de la méthode* (1637). But even without the added symbolism, the zealousness with which Voltaire assailed the reputation of his celebrated compatriot was apparent to all. So, too, were the potential ramifications of Voltaire's onslaught. Mme du Châtelet defended Voltaire in a February 1738 letter to a fellow Newtonian, Maupertuis – whom she fondly dubbed "Sir Isaac Maupertuis" – on the grounds that the French were blissfully ignorant of the fundamental flaws of Cartesian natural philosophy evident to everyone else in Europe. Her anonymous review of the *Elémens*, while mildly rebuking Voltaire for the harshness of his critique, nonetheless reiterated that France was "deplorably backward in its awareness" of the veritable revolution that had transformed natural philosophy; only recognition of the altered state of physics, she cautioned, would allow France to participate "in the progress which the Newtonian discoveries [would] make possible."[28]

The Newtonian revolution alluded to by du Châtelet had just then received its most spectacular public demonstration to date: confirmation of Newton's prediction in the *Principia* that the earth was shaped as an oblate spheroid, flattened at the poles. Based on the measurement of the length of a minute of an arc along the meridian, the confirmation was made by an expedition of the Académie des Sciences to Lapland, headed by Maupertuis. (Another expedition had been dispatched to Peru to initiate similar measurements along the equator.) Such measurements challenged the accuracy of the geodesic measurements carried out by the Director of the Paris Observatory, Jacques Cassini, on the basis of which he concluded (after Descartes) that the earth was a prolate spheroid. Maupertuis – the "flattener of the earth and of the Cassinis," as Voltaire would quip – returned to Paris in August 1737, and the preliminary reports of the expedition's findings instantly generated heated debate within the Académie as well as wide public interest. Maupertuis' published account, *Figure de la terre* ("Shape of the Earth"), followed in May 1738. Thus, the sensational news regarding the validation of Newton's theory was the talk of the town just as copies of the *Elémens* – as well as of Algarotti's *Newtonianismo* – were reaching Paris. The confluence of such accessible, pro-Newtonian literature created something of a siege mentality among Cartesians; virtually overnight, the grounds on which Descartes was defended were changed forever.

Fontenelle may well have been the first to articulate in public the new shift in balance between Newton and Descartes. Reading before the Académie des Sciences on April 16, 1738, his *éloge* of the Cartesian Joseph Saurin, who had died several months earlier, the Secretary cited Saurin's admonition that substituting Newton's attraction for Descartes' vortices would lead to the reinstatement of "the old darkness of Peripatetism, from which Heaven preserve us." Fontenelle then agonized: "Who would ever have thought it necessary to pray to Heaven to preserve Frenchmen from too favorable a bias for an incomprehensible system, they who love clarity so dearly, and for a System originating in a foreign land, they who have been charged with loving only that which is their own?"[29]

Reviewers and readers of the *Elémens* followed Fontenelle's lead, voicing concern over the conspicuous eclipse of Descartes' stature vis-à-vis Newton, as well as apprehension regarding the relevance of his work to contemporary natural philosophy. The abbé Granet, editor of the weekly *Réflexions sur les Ouvrages de Littérature*, for example, lamented that Newton had not lived before Descartes, "in which case Descartes' genius would have been appreciated and welcomed." Voltaire's former teacher, Father Tournemine, conceded in August 1738 that the *Elémens* succeeded admirably in rendering Newton intelligible; still, he disputed that Newton had dethroned

In this engraving by Johann Jakob Haid, Pierre-Louis Moreau de Maupertuis (1698–1759) is pictured "flattening" a terrestrial globe – an allusion to Maupertuis' confirmation of Newton's claim that the earth was an oblate spheroid, flattened at the poles. – Private Collection

Descartes. As a "compatriot" and "an old admirer of Descartes," Tournemine felt duty bound to offer an apologia for Descartes, which in retrospect is conspicuous for the historical terms in which it is couched. The state of philosophy when Descartes came on the scene must be remembered, the Jesuit maintained, in order to appreciate the strength of mind needed to overcome prejudices and establish fundamental principles of physics and metaphysics – not least of which was the true idea of God. Indeed, was it not Descartes who taught Newton how to philosophize? Tournemine went so far as to claim that "the system of Newton is almost the same as Descartes' system," except that by changing his terms, "Newton has substituted conjecture for certainty, embarrassment for clarity, and ignorance for evidence." He concluded with a protest against a Frenchman siding with an Englishman, suggesting that now that Voltaire had garnered fame as a poet as well as a philosopher, he would do well to apply himself to theology "without bias or prejudice."[30]

Voltaire responded with verve and gusto to these and similar charges of unpatriotic dalliances. A long letter to Maupertuis in October 1738 – published the following year – alluded to the controversy generated by the *Elémens* as well as to the bandying about of the names of Newton and Descartes as rallying cries. The Newtonian philosophy he expounded should not be made into a bone of contention between the English and the French, Voltaire insisted: it did not behoove an enlightened century to follow this or that philosophy on the basis of sectarianism. Privately, Voltaire was willing to concede that perhaps he ought to have shown more respect in his critique of Descartes and Malebranche; this, at least, was the conciliatory stance he took in a letter to Jean-Jacques Dortous de Mairan – whom he tried to convert to the Newtonian concept of action at a distance – and he certainly toned down his criticism beginning with the 1741 edition of the *Elémens*. Yet on substantive issues Voltaire remained unrepentant. He battled all charges of partisanship and treason; it was neither a "crime" to teach truths discovered in England nor the mark of a bad Frenchman to reject Cartesianism. (After all, the physicist Pierre Gassendi had been critical of Descartes, and he was no less of a Frenchman for it.) Elsewhere, Voltaire pointed out that both Descartes and Malebranche had criticized Aristotle; why, then, should *they* be regarded as "above reproach"? – especially as Descartes taught men to live by the injunction to follow truth alone.[31]

Jean-Jacques Dortous de Mairan (1678–1771) was among the first in France to replicate Newton's prismatic experiments and to simultaneously embrace significant elements of Newtonian science and modified elements of Cartesianism. – Smithsonian Institution Libraries, Washington, D.C.

Jean-Baptiste Rousseau, an inveterate enemy of Voltaire, made merry with the immediate flap created by the *Elémens*:

Rare mind, inventive Genius,
Which maintains for you alone
known Nature
Having no operative principle,
Except in the attraction maintained
by Newton;
V***, explain to us the attractive principle,
That caused to fall on your shoulders
These storms of blows from the Gaul,
Whose prize you received in real silver.[32]

The thinly veiled comparison of Voltaire to Judas Iscariot is indicative of the altered cultural landscape that had evolved by the 1740s. Voltaire, Maupertuis, and Algarotti had succeeded in awakening the French, as du Châtelet had so fervently wished. The scientific component of this new landscape was delineated by d'Alembert in his *Traité de dynamique* ("Treatise on Dynamics," 1743), where he characterized the Cartesians as "a sect that in truth is much weakened today." Fifteen years later, upon publication of a second edition, he rephrased his assessment: "a sect that in truth hardly exists today."[33] Although the editorial revision should not be interpreted as implying wholesale and unconditional conversion to Newtonianism by the middle of the eighteenth century, it does underscore that Cartesian natural philosophy had become seriously discredited. After the 1740s, only obscure savants or interlopers publicly and indiscriminately attacked Newtonian science; more established members of the scientific community opted for public silence or equivocation.

Consider the case of Dortous de Mairan. Following his death in 1771, a letter writer observed that the Académie des Sciences had "lost in him the last adherent of Descartes, whose chimerical systems have been entirely destroyed by the more luminous ones of Sir Isaac Newton. The Cartesian party in the Academy ha[d] become so weak, that M. de Mairan was too wise to defend the philosopher's visions on the subject of physics; he restrained himself to maintaining that Descartes had one of the most powerful minds which his age could boast; and on this point no one was disposed to contradict him."[34] Mairan, who was one of the first in France to replicate Newton's prismatic experiments, could more accurately be categorized as a Carto-Newtonian: a reflection of his simultaneous embrace of significant elements of Newtonian science and modified elements of Cartesianism (particularly a strong commitment to experimentalism). More relevant here is the historicist character of Mairan's reputed defense, which reflected a growing realization that contextualization alone could protect Descartes' fame and leave him relatively unscathed by comparisons to Newton.

Jean Le Rond d'Alembert (1717–1783) made significant contributions to mechanics and served as co-editor of the *Encyclopédie*. – Burndy Library

The twenty-three-year-old Anne-Robert-Jacques Turgot declaimed this burgeoning historical perspective in a speech he delivered at the Sorbonne in 1750. His oration, entitled "A Philosophical Review of the Successive Advances of the Human Mind," characterized Descartes as one who seemed "to wish to extinguish the torch of the sciences in order to relight it all on his own at the pure fire of reason." "Great Descartes," the orator continued, "if it was not always given to you to find the truth, at least you have destroyed the tyranny of error." Newton, in his turn, discovered "the key to the universe." He further "subjected the infinite to the calculus, has revealed the properties of light which in illuminating everything seemed to conceal itself, and has put into his balance the stars, the earth, and all the forces of nature." Elsewhere, Turgot asserted that Descartes "meditated and made a revolution." Yet no sooner had he shaken off the yoke of ancient authority than he fell back "on the very ideas of which he had been able to divest himself." Only respect prevented the future minister from comparing Descartes to Samson, "who, in pulling down the temple of Dagon, was crushed beneath the ruins." Ultimately, Turgot posited a necessary symbiosis "between these two mighty geniuses," analogous to a more general process in the growth of knowledge: great men open up new paths, lesser mortals clear them, "until a new great man rises up, who soars high above the point to which his predecessor had led the human race as that predecessor did above the point from which he set out."[35]

Turgot's marginalization of Descartes – now a historical figure relegated to a bygone century whose genius could not measure up to Newton's – was voiced in print in 1751, in d'Alembert's "Preliminary Discourse" to the *Encyclopédie*. While Descartes' contributions as a mathematician endured, d'Alembert maintained, in philosophy "everything remained to be done." Descartes "opened the way for us" and his "inventive genius" everywhere shone; but it was Newton who "gave philosophy a form which apparently it is to keep." D'Alembert continued: "Anything we could add to the praise of the great philosopher [Newton] would fall far short of the universal testimonial that is given today to his almost innumerable discoveries and to his genius, which was at the same time far-reaching, exact, and profound." As far as d'Alembert was concerned, no greater testimony of Newton's triumph existed than the disputes generated by his claims of discovery – "because at the beginning great men are accused of being mistaken, and at the end they are treated as plagiarists."[36]

Such studies in contrasts, modeled on the earlier efforts of Fontenelle and Voltaire, found parallels in works of philosophical fiction. These works disseminated even more widely the key tenets of the Newtonian gospel and embedded in the French national consciousness Newton's superiority over his French predecessor. The *Lettres Chinoises* ("Chinese Letters," 1741) of the Marquis d'Argens is a case in point. This purported correspondence between a Chinese visitor to Paris and a friend in China offers critical insights into the political, social, and cultural climate of the early eighteenth century. One letter, which took up the state of philosophy, described how within the previous two decades

> the Order and Rule of the Universe [had] changed two or three times. Descartes made an infinite Number of Worlds swim in a subtle Matter; and all those Worlds plentifully furnished, as ours is, with Sun, Moon, and Stars, were invironed each with a Vortex of Matter extremely thin and light, which ran all as fast as they could into an immense and infinite Fluid. But at present all this is changed. An Englishman has by his Omnipotence destroyed all those Vortexes, he has annihilated the Fluid which kept them up, and has established an immense Vacuum, in which he makes the Stars roll at their Ease, without any thing to incommode their Course.

To forestall the objection that "without a particular Cause, it was impossible that a Body could always preserve its circular Motion," this "new Creator, as it were, of the Universe" immediately endowed matter with "a new Quality, called *Attraction*, by which the Stars have a continual Tendency towards the Centre of their Motion. He ordered all Bodies to attract each other mutually, according to their Size, or to use his own Terms, according to the *Inversion of their Square of Distance*. From that time all Bodies gravitated to one another, and mutually attracted each other by the inviolable and unalterable Laws of Attraction."[37]

Along with Voltaire, d'Alembert, and most of the other *philosophes*, *Encyclopédie* editor Denis Diderot (1713–1784) was a harbinger of the "Anglomanie" that overtook France by the mid-eighteenth century. – Burndy Library

In his 1748 salacious novel *Les bijoux indiscrets* ("The Indiscreet Jewels"), Denis Diderot took a more engaging approach. The story focuses on a Congolese Sultan who is given a magic ring with the power to make women's private parts speak. A scene in which the local Academy of Sciences investigates the properties of the ring enables Diderot to spin an entertaining account of the schism between a party of Vorticists, founded by "a skilful geometer and great physicist" by the name of Olibrio, and an Attractionist party, founded by Circino, "an able physicist and great geometer." The principles of the Vorticists have, at first glance, a seductive simplicity. They satisfy the principal phenomena on the whole, but they contradict themselves in the details. As for Circino, he appears to base his ideas on an absurdity; but it is only this first step that works against him. The minute details that ruin Olibrio's system serve to establish Circino's. He follows a path that is obscure at the beginning, but which becomes more and more clear as one advances. Olibrio's, on the contrary, clear at the beginning, heads over toward obscurity. Olibrio's philosophy demands less study than understanding, whereas one cannot be a disciple of the other without much understanding and study. No preparation is necessary to enter into Olibrio's school. Anyone can hold its key. Circino's is accessible only to the finest geometers. Olibrio's vortices are within the reach of all minds. Circino's central arguments are for first-rate algebraists alone. There will therefore always be one hundred Vorticists for every Attractionist; and one Attractionist will always be worth one hundred Vorticists.[38]

Diderot and d'Alembert, along with Voltaire and most other *philosophes*, were the harbingers of the "Anglomanie" that overtook France by the mid-eighteenth century. Historian Edward Gibbon noted the prevailing mood upon visiting Paris in 1763: English opinions, English fashions, and even English games, he found, "were adopted in France: a ray of national glory illuminated each individual, and every Englishman was supposed to be born a patriot and a philosopher."[39] But Anglomania also rankled many Frenchmen. As Newton became the epitome of English greatness, a growing need arose in France to secure Descartes' stature, which had undergone the marginalization discussed above. The result was Descartes' induction into the "secular cult of great men," celebrated

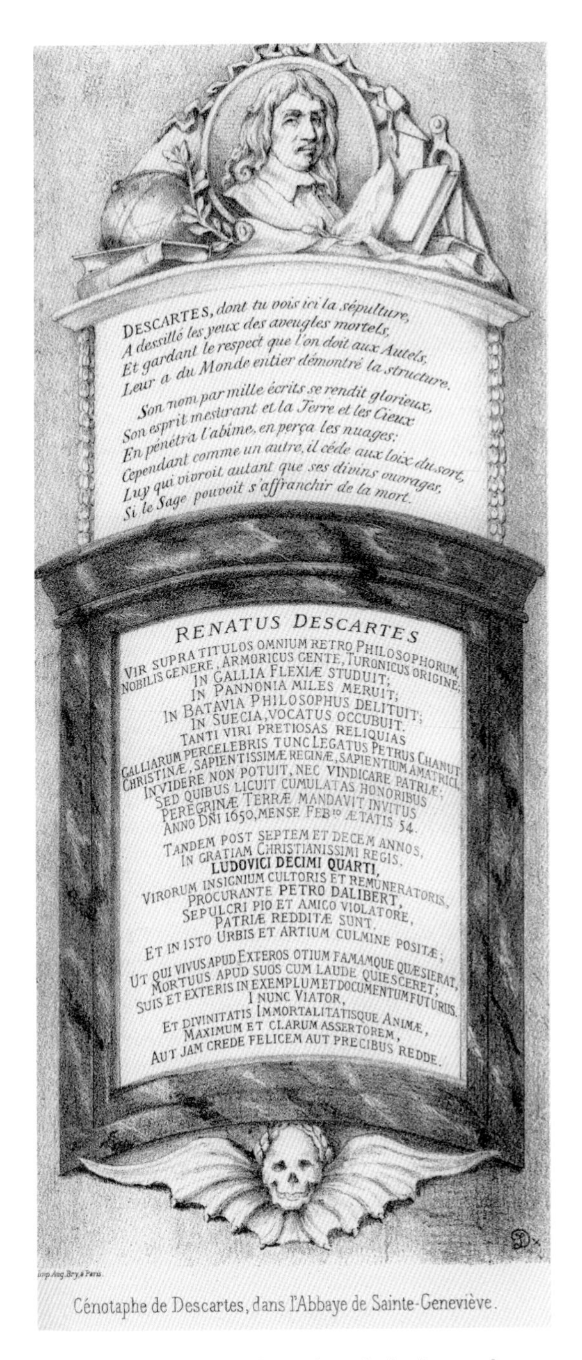

Descartes died in Sweden, where he had served as tutor to Queen Christina, in 1650. His remains were interred in the Church of Ste. Geneviève-du-Mont in Paris in 1667. This reproduction of the monument there appeared in A. Boulay de La Meurthe's "Monuments Funéraires de Descartes," published in the *Mémoires de la Société Archéologique de Touraine* 23 (1873). - Private Collection

Monument de Descartes par Sergel dans l'Eglise Adolfe-Frederick. *(Stockolm).*

In 1777, Gustav III, the francophile king of Sweden, instructed the royal sculptor Johann Tobias Sergel to create a monument to Descartes, for the Adolf Fredrik Church in Stockholm. From A. Boulay de La Meurthe's "Monuments Funéraires de Descartes" (1873). – Private Collection

annually by the Académie Française. Charles-Irénée de Castel, abbé de Saint Pierre, was instrumental in conferring such "academic immortality" on Descartes. In the wake of Voltaire's *Elémens*, the octogenarian academician published *Discours sur les différences du grand homme et de l'homme illustre* ("Discourse on the Difference Between a Great Man and an Illustrious Man"). Here Saint Pierre pronounced that however wondrous the contribution of an illustrious man, he could still lack the "virtues" that make a man "great." Using precisely such a distinction, he argued that it was Descartes' cast of mind and contribution to humanity – rather than his mathematical prowess – that warranted his inclusion among "the greatest men that ever existed." A similar zeal to enshrine Descartes inspired the Académie Française in 1755 to award the annual prize for eloquence to the Jesuit Antoine Guénard, who responded to that year's theme – "what constitutes philosophical spirit" – with the demonstration that the answer resides in thinking for oneself; Descartes, no surprise, is thrust forward as the exemplar of precisely such a philosophical spirit.[40]

A decade later, the Académie sponsored a competition for an *éloge* to Descartes. The contestants – who included such celebrated literary figures as Antoine-Léonard Thomas, Louis-Sébastien Mercier, and Claire-Marie de Saint-Charmond, marquise de Vieuville – outdid each other with effusive panegyrics in praise of the noble spirit of this great Frenchman, his singular humanity, and his deliverance of mankind from ignorance, prejudice, and servility to authority. If the contestants could neither ignore Descartes' scientific errors nor deny Newton's superiority, they could nevertheless insist that even Descartes' errors were sublime. Thus, Mercier's florid eulogy breaks off to contrast the two geniuses – mostly to Newton's advantage – only to speculate that were Descartes to return to this world he, as the friend of truth, would acknowledge his defeat – which would in no way diminish the Frenchman's greatness. If anything, it was Newton's name that derived brilliance from its association with that of Descartes. Thomas, the co-winner of the prize, actually structured his eulogy around such a contrast. Opening with the defiant statement that he would proclaim Descartes' *éloge* from the feet of Newton's statue – would even make Newton himself praise his great master – Thomas proceeds to manipulate Newton as a foil for Descartes. Newton discovered many truths, but Descartes had discovered the route to these truths. Both Descartes and Newton were sublime geometers, although Descartes never utilized geometry to the same extent as Newton; Newton was a genius, but so was Descartes, and more original to boot, even if his genius sometimes misled him. And so on.

Couched in these historical terms, even Voltaire was willing to acquiesce to Descartes' "greatness." "One no longer reads Descartes," he wrote Thomas, "but one reads your éloge." Voltaire admired particularly Thomas's ability "to separate the genius of Descartes from his chimeras." For the eulogists of the eighteenth century, however, even the chimeras served to reconstitute and refurbish the image of Descartes: "Let's respect the errors that marked the seal of the genius," another eulogist, Couanier Deslandes, proclaimed; "The vortices honor his reason and their ruins are sacred for us." On the ruins of these ideas, Descartes was baptized "father of the academicians, since he was the incontestable founder of the philosophical spirit." [41]

This paternity would eventually lead in the nineteenth century to the conviction that "Descartes is France." [42] For now, however, his apotheosis was shadowed, even in France – as we shall see in chapter 7 – by the looming figure of Newton and even of more recent celebrated Frenchmen, such as Buffon. Only in 1775 did Count Charles-Claude de Flahaut de la Billarderie d'Angiviller, director of Louis XVI's buildings, decide to include a statue of Descartes in a projected gallery of great Frenchmen, entrusting Augustin Pajou with the task. For whatever reason, the sculptor appears not to have been particularly inspired by his subject. In the words of one contemporary, Descartes "looks far-off and dreamy, rather than impressing us with the powerful thoughts of a philosopher creating the world in his imagination." Certainly, nothing of the power so evident in Pajou's contemporaneous statues of Buffon and Pascal carried over into his Descartes. The philosopher simply stands there, an empty scroll in his left hand, a stack of books behind him and a pair of compasses by his feet. The only indication that Pajou kept Descartes in mind while working is the likeness to the famous portrait by Frans Hals he had imprinted on his marble. [43] In marked contrast, Robert-Guillaume Dardel's 1782 personification of the Frenchman captures the philosopher's spirit. His "Descartes piercing the darkness of ignorance" depicts Descartes attempting to extricate himself from a dense,

Robert Guillaume Dardell's "Descartes Piercing the Darkness of Ignorance" (1782) depicts the French philosopher's supreme effort to extricate himself from a dense, ensnaring mass of clouds, an effort in which Descartes is distracted by the bright rays of the sun beginning to break through. - By kind permission of the trustees of The Wallace Collection, London

ensnaring mass of clouds. Descartes' supreme effort to free himself – one foot is still trapped in the clouds – appears to be interrupted, his attention suddenly distracted by the bright rays of the sun beginning to break through. Descartes has nearly succeeded in vanquishing ignorance and heralding a new enlightenment, Dardel seems to suggest, but then a momentary lapse robs him of the triumph within reach. Dardel's message seemed to be reinforced a decade later when he celebrated in stone the person who did effect the emancipation that eluded Descartes – Isaac Newton.

The crucial role that nationalistic sentiments played in the reception of Newtonianism must be understood. In 1728, a Swiss traveler to England extolled, much as Voltaire would a few years later, the "Freedom of Thoughts and

In 1707, in his *La verité recherché par les philosophes* (left), Bernard Picart depicted Lady Philosophy guiding Descartes – at the head of a group of philosophers – toward Truth. An enterprising English printer reproduced the engraving (right), replacing Descartes' portrait with that of Newton, and altering the accompanying text to indicate that Newtonian gravity had wiped out Cartesian vortices. – The Metropolitan Museum of Art (left); Private Collection (right)

Sentiments" that permitted Englishmen to distinguish themselves in the domain of learning. Simultaneously, however, he also broadcast a widespread charge leveled against the English: they habitually appropriated all new ideas, denying the contribution of others. "This is a Pretension so like to disturb *Parnassus*, and to stir up Disputes," the visitor warned, adding that he had already picked up the sound of the charge "and the *Literati* running to Arms."[44] A wonderful illustration of this allegation can be found in an early eighteenth-century plagiary of a French print.

In 1707, Bernard Picart was commissioned to decorate a thesis by an obscure student at the Jesuit Collège Louis-le-Grand in Paris. Entitled "La vérité recherchée par les philosophes" ("Truth Sought by Philosophers"), the print depicts Lady Philosophy leading Descartes – who is followed by Aristotle, Plato,

Socrates, and Zeno – toward Truth. As the text underneath explains, Philosophy, holding Descartes' right hand, points to the battle just ahead of them, in which Father Time dispels the clouds obscuring Truth and helps Minerva defeat ignorance. A few of the rays emanating from the sun held by Truth illuminate the ancient philosophers; most of the brilliance, however, is reserved for Descartes, embodying the lesson that those who come last perfect the knowledge and the discoveries of their predecessors. In Descartes' left hand is also a scroll upon which his account of vortices is inscribed with such precision, the text states, that "it would be inexcusable to mistake it." In a fascinating act of plagiarism, an anonymous Englishman reproduced the image and translated the accompanying text – with minor (but extremely significant) changes: Descartes' portrait is replaced by Newton's; so, too, the scroll is blank, the text beneath indicating that gravity has wiped out vortices. Most significant, the light emanating from Truth, which in the original print illuminated in its path only Descartes, now extends over the entire print – an incomparable representation of Newton's triumph.[45]

The nationalistic overtones of the scientific debates reverberated for the next half century and more. In his preliminary discourse to the *Encyclopédie* (1751), d'Alembert praised Maupertuis for courageously holding to the principle "that one could be a good citizen without blindly adopting the physics of one's country." Maupertuis himself conceded that little glory was to be gained in presenting one's compatriots with the discoveries made by others some fifty years earlier – as his own experience had demonstrated. Later in the century, however, once French glory had been restored, it became possible to view past events more dispassionately. In his 1774 *éloge* of La Condamine – who had participated in the expedition to Peru – Condorcet noted that nationalistic sentiments had interfered with scientific impartiality earlier in the century; yet the French, he claimed, had outgrown this bias and regained the initiative: "the system of the world, that most imposing monument of forces and the greatness of the human mind, built by Newton on immovable foundations, had been raised by the genius of his successors to a height" scarcely imaginable. Two decades later, the chemist Antoine Laurent Lavoisier embraced precisely such sentiments in his last-ditch effort to preserve the Académie des Sciences from the iconoclasm of the French Revolution. Thanks to the Académie, he informed the Convention, French science had surpassed English science: "We would be dishonored if our scientists were reduced to carrying their talents and our shame to foreign shores," he exclaimed, especially since it was French savants who "have enlarged the mass of knowledge that we owe to Newton's genius."[46]

This portrait by Jacques Louis David of the French chemist Antoine Laurent Lavoisier (1743–1794) and his wife was commissioned by Mme Lavoisier in 1787, the centenary of the *Principia*, and the year in which Lavoisier's revolutionary new nomenclature of chemistry was published. Completed the following year, David's masterpiece celebrated Lavoisier – best known for his pioneering studies of oxygen, gunpowder, and the chemical composition of water – as the new hero of science, who revolutionized chemistry by establishing it on solid experimental and rational principles, akin to the manner in which Newton revolutionized physics. – The Metropolitan Museum of Art

5

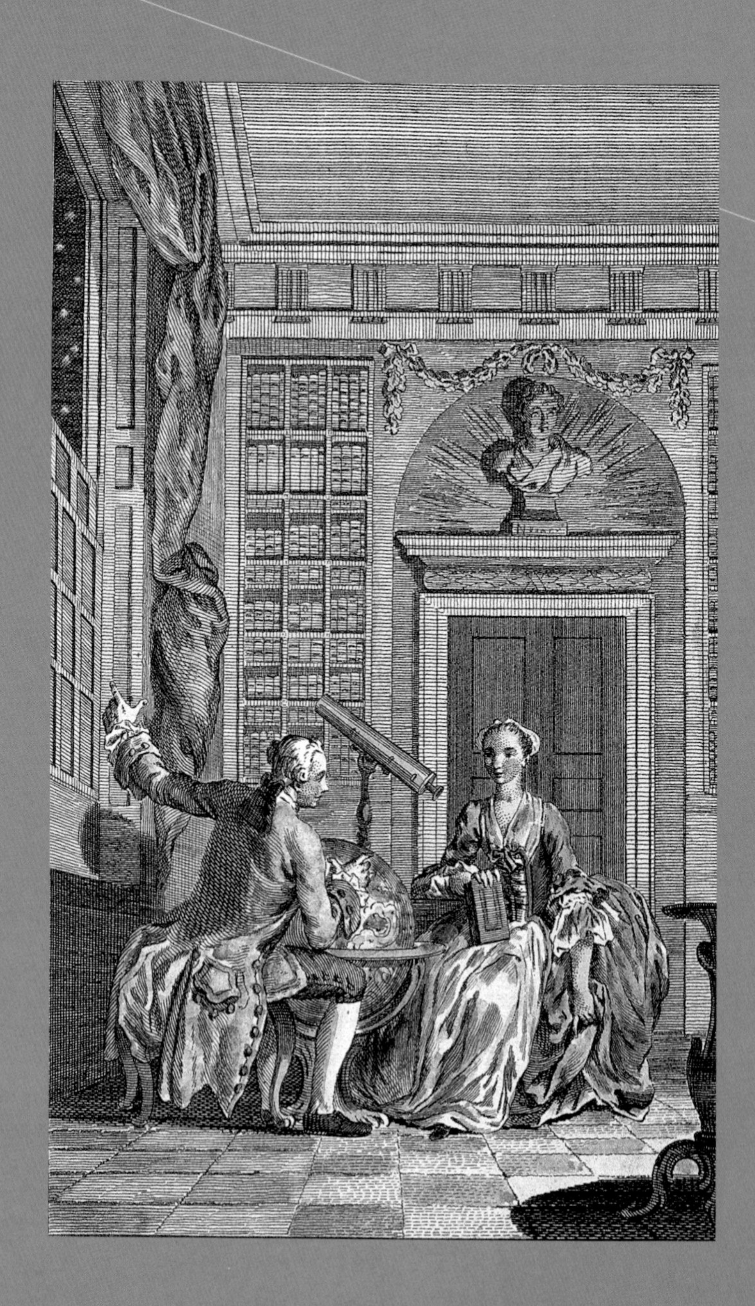

Benjamin Martin was an instrument maker, itinerant lecturer, and prolific author of scientific works that popularized Newton's philosophy. This illustration is from his *The Young Gentleman and Lady's Philosophy* (London, 1755), published as Part 1 of *The General Magazine of Arts and Sciences* (1755–65). – Courtesy of Adler Planetarium & Astronomy Museum, Chicago, Illinois

NEWTONIAN WOMEN

One of the most interesting manifestations of the struggle for hegemony between the Cartesians, Newtonians, and Leibnizians was the campaign for the allegiance of women, who by the late seventeenth century had become consumers of all matters scientific. Descartes had been instrumental in this reorientation of women's interests, as he was perceived early on as the philosopher of women – a perception he himself fostered. Not only had he composed his *Discours de la méthode* ("Discourse on Method," 1637) in French rather than Latin, thereby ensuring its accessibility to a broadly nonacademic audience, but he had simplified his proof of God's existence, he told a correspondent, so that even women "could understand something." His 1644 dedication of the *Principia philosophiae* to Princess Elizabeth of Bohemia gave public voice to this new inclusiveness; after complimenting his royal correspondent for escaping the common fate of women – to be condemned to ignorance – he singled her out as the only person who had "completely understood" his philosophy.

Diamante Medaglia Faini (1724–1770) was a celebrated Italian poet and member of the Accademia degli Unanimi in Salò. In 1764, she publicly renounced poetry in favor of mathematics, an act commemorated in the allegorical frontispiece to her *Versi e prose* (Salò, 1774), shown here. – Private Collection

Descartes' radical philosophy had particular relevance for women by virtue of its grounding in introspection rather than in erudition, and the fundamental distinction between mind and body that Descartes posited as its underlying principle. The implications were far-reaching; at once, the female constitution became, for all intents and purposes, "irrelevant" to the capacity to reason, thus doing away with the traditional impediment to the pursuit of abstract sciences by women. In a similar manner, the subjectivity of Descartes' philosophy rendered formal schooling – to which women were denied access – suddenly beside the point. Not surprisingly, not a few women (and the occasional man), taking their cue from Descartes, were emboldened to advocate a reorientation of women's education in line with the new philosophy. As Giuseppa-Eleonora Barbapiccola announced in the preface to her 1722 translation of the *Principia philosophiae* into Italian, her goal was to make Descartes accessible particularly to women, attributing to Descartes himself the observation that women were "more apt at philosophy than men."[1]

Descartes thus became the "ladies' philosopher." The radical new philosophy that swept away all past learning (and errors) concurrently consolidated women's capacity for reason with their fundamental piety. The vehemence of the orthodox intelligentsia in combating Cartesianism also lent the new philosophy the "irresistibility of the forbidden fruit."[2] Perhaps the writer who did the most to propagate a Cartesian feminine philosophy was François Poulain de la Barre, whose *De l'égalité des deux sexes* ("The Equality of Sexes," 1673) introduced the slogan "the mind has no sex." To the twenty-six-year-old graduate of the Sorbonne, the advocacy of women was part of a broader radical program, aimed at applying Cartesian principles to the reform of politics and society. In good Cartesian fashion, de la Barre argued that the only prerequisite for the pursuit of science was the acquisition of a true method and the possession of reliable senses. Since Descartes denied any distinction between male and female senses, both sexes could rely on their senses with equal confidence. Thus, it followed – for de la Barre if not for most of his contemporaries – that women, if equipped with a true method (Descartes', of course), could become as capable and as adept as men in contributing to whatever art or science they set their sights on; in other words, no domains were exempt from the female reach.[3]

The radical worldview took root. Significant numbers of learned women embraced Cartesian ideas, providing an important source of support at a time when Descartes' philosophy faced persecution both by the institutions of higher learning and the Church. Nevertheless, since only a few women published or left other written records of their learned pursuits, their broad success in mastering philosophical and scientific knowledge becomes difficult to gauge. What is clear, however, is the elevation of Cartesianism as an important topic of conversation in various salons – such as those of Madeleine de Scudéry, the Marquise de Sablé, and Marguerite Hesseub de la Sablière. Other women – such as Catherine Descartes, the philosopher's niece, and Anne de la Vigne – demonstrated their Cartesian leanings in their literary efforts. Most extraordinary of all, a few women began to participate – albeit primarily as spectators – in the public lectures and demonstrations offered by such Cartesian propagandists as Pierre Sylvain Régis in Toulouse and Jacques Rohault in Paris. At least two women who attended Rohault's sessions, Mmes de Guederville and de Bonneveau, were inspired to establish their own private experimental and philosophical soirées,[4] a trend that proliferated in subsequent decades.

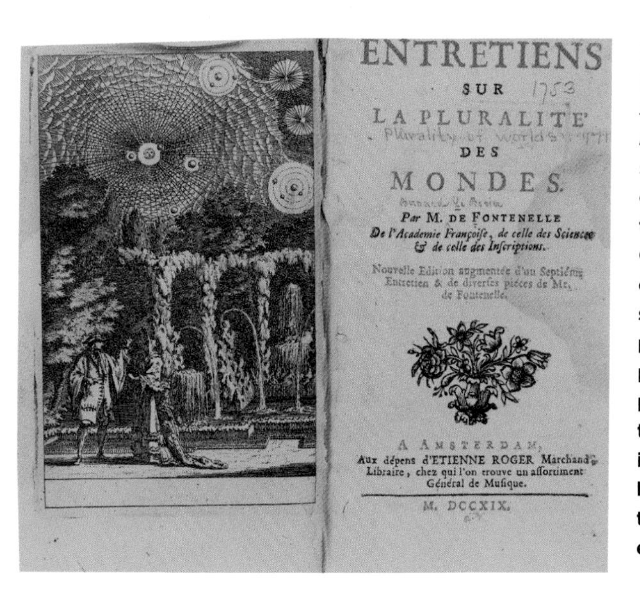
ENTRETIENS
SUR
LA PLURALITE'
DES
MONDES.
Par M. DE FONTENELLE
De l'Academie Françoise, de celle des Sciences & de celle des Inscriptions.
Nouvelle Edition augmentée d'un Septiéme Entretien & de diverses piéces de Mr. de Fontenelle.
A AMSTERDAM,
Aux dépens d'ETIENNE ROGER Marchand Libraire, chez qui l'on trouve un assortiment Général de Musique.
M. DCCXIX.

In Bernard Le Bovier de Fontenelle's *Entretiens sur la pluralité des mondes*, a learned savant initiates an eager marquise into the mysteries of Cartesian cosmology over the course of several nights. First published in 1686, it proved immensely popular, with nearly thirty editions appearing in the following half century, including this 1719 Amsterdam edition. – NYPL-SIBL

In no small part thanks to the efforts of Fontenelle, the vogue spread. His popularization of Cartesian cosmology in the 1686 *Entretiens sur la pluralité des mondes* ("Conversations on the Plurality of Worlds") became so influential that, thirty-five years later, the young Voltaire still acknowledged Fontenelle's sway in a facetious verse letter to the Secretary: the women he kept company with had been "spoiled" by reading the *Entretiens*; too bad they hadn't been spoiled by Fontenelle's pastorals instead, Voltaire lamented, "for we would much rather have seen them shepherdesses than philosophers." As it was, since the taste of men "is regulated by theirs, love for them has made us all turn natural philosophers."[5] Fontenelle himself turned commentator on the phenomenon in his *éloge* of Louis Carré. This mathematician-priest burned his letters in order to avoid scandal, Fontenelle explained, because so many of them came from women seeking out scientific tuition and advice. In 1713, the English natural philosopher John Theophilus Desaguliers boasted that "the Newtonian philosophy was generally received among persons of all ranks and professions, and even among the ladies by the help of experiments." By the late 1750s it appeared as if science and courtship had joined ranks, sufficiently at least for Oliver Goldsmith to propose that he "who would court a lady must be capable of discussing Newton and Locke." Across the Channel, an observer of the French scene in the 1770s attributed the fashionableness of mathematics to the "charlatanism" of Maupertuis, as a consequence of which "women were all bit by a passion for the society of geometricians, and it was the height of *bon ton* to have them among the parties at all the suppers."[6]

Women's enthusiasm for science became such that contemporaries routinely commented on its effects on the reorientation of the language and content of science. Voltaire spoke too much as a philosopher, complained a somewhat hostile reviewer of the *Lettres philosophiques*. He reasoned with Newton and Locke, whereas the fair sex – which made up the greater part of the reading public – would gladly have dispensed with such erudition in favor of a charmer who walked his readers on the moon and eased the dryness of the subject matter with pleasant fiction (as Fontenelle did). Much the same stance was taken by Diderot several decades later: "women accustom us to discuss with charm and clearness the dryest and thorniest subjects," he noted. "We talk to them unceasingly: we listen to them: we are afraid of tiring or boring them. Hence we develop a particular method of explaining ourselves easily that passes from conversation into style."[7]

Margaret Bryan ran a girls' boarding school, as well as offered private tutorials in science. She is seen here, in the company of her two daughters, in the frontispiece to *A Compendious System of Astronomy* (London, 1797), a collection of her lectures. – NYPL–Pforzheimer Collection

The enthusiasm with which increasing numbers of women embraced the new philosophy did little to change the prevailing prejudice – even among Cartesians – against their active participation in abstract learning beyond an elementary level. The Jesuit Pierre Le Moyne, for example, who boldly stated in 1647 that women were as capable as men of comprehending speculative philosophy, hastened to add that he should not be construed as advocating university education for women or, in his words, that spheres and astrolabes be put into their hands rather than wool and needle.[8]

Descartes' great philosophical heir, Nicolas Malebranche, was positively hostile in his conviction of the unsuitability of the delicate female brain to grasp abstract learning. Women were perfectly competent to handle matters pertaining to taste, he pronounced; "but normally they are incapable of penetrating to truths that are slightly difficult to discover. Everything abstract is incomprehensible to them. They cannot use their imagination for working out complex and tangled questions. They consider only the surface of things, and their imagination has insufficient strength and insight to pierce it to the heart."[9]

These negative reactions to the growing (and public) involvement of women in scientific studies in the seventeenth century only strengthened in subsequent decades as the movement broadened. Philosophers and men of letters were nearly unanimous in their judgment that women could not, and should not, participate equally in the scientific enterprise. Jean-Jacques Rousseau, the great apostle of the emotions and the favorite of women, is a case in point. His phenomenally popular *Emile* (1762) came close to advocating that women be kept in a state of blissful ignorance. He desisted, it seems, mostly because in an age of philosophy, their domestic duties required familiarity with the sort of learning they might encounter in social situations; so, too, some learning would help safeguard their virtue. Yet for Rousseau, women were endowed with neither the capacity nor the need to wield their minds abstractly. Exceptions existed, he conceded; but in line with common stereotypes, he protested that erudition had turned such women into men; he would want such a prodigy as neither a friend nor a mistress. Thus, for all his warm embrace by women, and his reputation as the philosopher of feeling, Rousseau maintained that "the search for abstract and speculative truths, for principles and axioms in science, for all that tends to wide generalization, is beyond a woman's grasp; their studies should be thoroughly practical." And more idiomatically: "Woman has more wit, man more genius; woman observes, man reasons."[10]

In his *A.L.M. Naturlehre* (Halle, 1740–43), Johann Gottlob Krüger, a professor of medicine and philosophy at the University of Halle, glorified the study of nature and the pursuit of its laws. This is the frontispiece to volume 1. – Smithsonian Institution Libraries, Washington, D.C.

At the opposite end of the philosophical spectrum stood Immanuel Kant, who nonetheless preached an identical message when it came to women. Though nature endowed men and women equally with understanding, the fair sex was allotted a "*beautiful understanding*," their male counterparts a "*deep understanding*" – the latter expression signifying for Kant "identity with the sublime." The aspirations for "deep meditation and a long-sustained reflection are noble but difficult," Kant continued, and utterly unbecoming to a woman. Abstruse learning would turn her into "an object of cold admiration," at the same time obliterating the charms proper to her sex. An erudite Grecian, like Mme Dacier, or a profound physicist, like Mme du Châtelet, Kant was convinced, "might as well even have a beard; for perhaps that would express more obviously the mien of profundity for which she strives." Small wonder, then, that Kant's conclusion regarding the kind of learning women should pursue was conspicuously bereft of abstract knowledge:

> Beautiful understanding relinquishes to the diligent, fundamental, and deep understanding abstract speculations or branches of knowledge useful but dry. A woman therefore will learn no geometry; of the principles of sufficient reason or the monads she will know only so much as is needed to perceive the salt in a satire which the insipid grubs of our sex have censured. The fair can leave Descartes his vortices to whirl forever without troubling themselves about them, even though the suave Fontenelle wished to afford them company among the planets; and the attraction of their charms loses none of its strength even if they know nothing of what Algarotti has taken the trouble to sketch out for their benefit about the gravitational attraction of matter according to Newton.[11]

Those portraying themselves as "friends" of women expressed the same dismissive attitude. The editor of the short-lived *Bibliothèques des Femmes*, who battled the common prejudice that denied women access to education and to the professions, saw no contradiction in simultaneously encouraging women to embrace the sciences and affirming it well "beyond the female character to match Euclid, Kepler, or Newton." Taking a similar position, the first editor of the *Journal des Dames* regretted the sight of women struggling "with the thorny problems of science, forever inaccessible to [their] vain curiosity." They should garner instead, he suggested, "the roses that our poets produced" especially for their pleasure.[12]

The reluctance to encourage powerful female intellects derived, in no small part, from male pride. Another "friend" of women, Pierre-Joseph Boudier de Villemert, spiritedly advocated the cultivation of women's minds – albeit primarily for the benefit of men who would reap the fruit of their availability for cultivated conversation – but with the proviso that "abstruse sciences and thorny researches," which "oppress their minds, and blunt that ingenuity," be excluded. Once again we hear that Mmes Dacier and du Châtelet are worthy of admiration, not imitation. Women would do better to cultivate natural history, grounded as it is in curiosity and the imagination; the more dizzying laws of physics should be left to men. By abiding by such distinctions, women might "exercise their minds," even "surpass us, without humbling us."[13]

The fear of male humiliation if women set out in pursuit of abstruse learning is voiced again and again. James Boswell confided to Samuel Johnson that Belle de Zuylen, whom he considered a possible match, was "a charming creature[;] but she is a savante and a bel esprit, and has published some things. She is much my superior. One does not like that."[14] Even more instructive is the account of Charles de Brosses' 1739 encounter with Maria Gaetana Agnesi. After recounting how he and his friends reneged on an invitation to visit the salon of the learned Clelia Borromeo, the future French magistrate gives voice to even greater apprehension at the prospect of meeting the prodigious Agnesi:

> This evening it will be even worse: We have to go to a meeting with Signora Agnesi, twenty years of age, who is a walking dictionary of all languages and who, not content with knowing all the oriental languages, gives out that she will defend a thesis against all comers about any science whatever.... In faith, I don't the least desire to go; she knows too much for me. Our only recourse is to let loose upon her Loppin [a fellow traveler] for geometry, in which our *Virtuosa* principally excels.

Evidently, de Brosses' curiosity got the better of him, and, as we shall see below, he kept the appointment.[15]

Fear of intellectual emasculation, central to the male objection to scientific erudition in women, was matched by considerable female disapproval, often for similar reasons. Mary Wollstonecraft's take on the subject is telling. In her *Vindication of the Rights of Woman* (1792), this celebrated advocate of women nonetheless maintained that the singularity of the few ought not necessarily encourage emulation: just as Newton "was probably a being of a superior order accidentally caged in a human body," so, she mused, "the few extraordinary women who have rushed in eccentrical directions out of the orbit prescribed to their sex, were male spirits, confined by mistake in female frames."[16]

Clearly, then, a handful of prodigies neither challenged common prejudice nor posed a substantial threat – and precisely for this reason were generally tolerated, even touted. Indeed, remarkable women became something like freaks of nature, specimens worthy of cultivation and display just like the other natural rarities collected in the early-modern period. Many celebrated women, including Newtonian women, owed their opportunities to pursue science to the precocity they displayed as girls, and to the willingness – even desire – of family members to make public spectacles of them. Their singularity was thereby turned into a commodity to be exploited to the hilt. And yet it was this very fascination and exploitation that afforded scores of women in the eighteenth century opportunities to study and practice the sciences in ways that would otherwise have been unimaginable.

Most spectacular in this regard, perhaps, is the career of Laura Bassi. Born in 1711, Bassi was the only surviving child of a Bolognese lawyer, who consequently bestowed on her the sort of learned attention normally reserved for male offspring. At five, she began studying Latin, French, and arithmetic; philosophy was added at thirteen. Her exceptional mind as well as her skills in disputation soon became renowned, and Bassi exhibited both in public. Her fame was enshrined on May 12, 1732, when the University of Bologna conferred on her a doctorate in philosophy. Through the influence of Cardinal Lambertini (the future Pope Benedict XIV), Bassi was appointed to a philosophy lectureship – but because of her sex, she rarely taught on university premises. Instead, students flocked to the private lectures in experimental physics that she offered at home. Such lectures commenced, however, only after her marriage to Giuseppe Veratti in 1738. In the three years leading up to her marriage, Bassi studied mathematics with Gabrielle Manfredi while collaborating with several members of the Institute of Sciences in Bologna, to which she belonged. The close association of a single young woman with these men appears to have generated gossip, and Bassi realized that if she was to carry on her scientific work, her only option was marriage. As she intimated to a friend, "I have chosen a person who walks my path in the arts and who, through long experience, I was certain would not impede me from following mine."

A child prodigy, Laura Bassi (1711–1778) was the second woman to receive a doctorate from an Italian university (1732). She was then appointed to a philosophy lectureship at the University of Bologna and became a member of the Bologna Institute of Sciences. – By permission of the Ministero per i Beni e le Attività Culturali

The theses Bassi defended in 1732 as part of the requirements for her philosophy degree reflected the Cartesian leanings of her teacher, Gaetano Tacconi. But when she began to lecture on her own later that year, her marked preference for Newtonian natural philosophy emerged. Indeed, in her first lecture she propounded Newton's credo that the philosopher's duty was "to deduce the laws that governed nature from phenomena that could be observed experimentally." She then proceeded to treat Newton's laws of motion. For the following four decades, she experimented and lectured continuously, and these mostly public events became an important platform for the diffusion of Newtonianism in Bologna and beyond. Many young scholars derived their initial scientific training – as well as their allegiance to Newtonianism – from her tutelage; a case in point is Bassi's cousin Lazzaro Spallanzani, who came to Bologna to study law but was quickly converted to natural philosophy.[17]

Bassi was not the first woman to receive a university degree. Elena Lucrezia Cornaro Piscopia was so honored by the University of Padua as early as 1678, with an estimated crowd of 20,000 spectators witnessing the spectacular and unprecedented event in the town's cathedral. At the insistence of her father – who was determined to display the extraordinary talents of his daughter – Piscopia had gone to Padua to study and later to apply for the degree. And although for the remaining six years of her life she continued to amaze visitors with her multifaceted erudition in languages, philosophy, and the mathematical sciences, her focus increasingly turned to works of devotion and charity.

A similar trajectory was followed in the career of Maria Gaetana Agnesi. She owed her extraordinary education to the forceful social aspirations of her father, who sought to catapult himself into the Milanese patriciate partly by putting his prodigious daughter on display in the salon he established at his home. In 1727, at the age of nine, Agnesi declaimed from memory a Latin oration – apparently penned by her tutor – against the common prejudice preventing women from gaining access to the study of arts and sciences; the oration was published two years later in a volume on the topic sponsored by the Accademia dei Ricovrati (Academy of the Sheltered Ones). Agnesi completed an intensive course of philosophical and theological study in 1738, and published a list of 191 theses in logic, physics, and metaphysics that she had defended over the previous few years in her father's salon. Though the Cartesian framework informed the general philosophical view in this publication, Agnesi availed herself repeatedly of Newtonian science; she demonstrated the applicability of Newton's laws of motion to several areas of natural philosophy, and spoke approvingly of his theory of light and colors.[18]

The following year, Charles de Brosses visited the Agnesi salon, and his description of the conduct of the person he described as "something more stupendous than the cathedral of Milan" is poignant:

I was brought into a large and fine room, where I found about thirty people from all the countries of Europe, ranged in a circle, and Mlle Agnesi, all alone with her little sister, seated on a sofa. She is a girl eighteen to twenty years of age, neither ugly nor pretty, with a very simple and very sweet manner.... Count Belloni, who took me, wished to make a kind of public show. He began with a fine harangue in Latin to this young girl, so as to be understood by everybody. She answered him very well; after that they entered into a disputation, in the same language, on the origin of fountains and on the causes of the ebb and flow which are observed in some of them, like the tides of the sea. She spoke like an angel on this subject; I have never heard anything that gave me such pleasure.

The private and public lectures in philosophy and experimental physics that Bassi delivered for some four decades were instrumental in disseminating Newtonian science in Bologna and beyond. Her fame – and singularity – drew considerable audiences to her public lectures. – By permission of the Ministero per i Beni e le Attività Culturali

Another child prodigy, Maria Gaetana Agnesi (1718–1799) owed her extraordinary education to the forceful social aspirations of her father, who sought to catapult himself into the Milanese patriciate partly by putting his daughter on display. – Burndy Library

De Brosses also recounted that Agnesi was "much attached to the philosophy of Newton, and it is marvellous to see a person of her age so conversant with such abstract subjects."[19] Yet, like Piscopia half a century earlier, Agnesi's heart appears not to have fixed on the sciences. A deeply devout person, she expressed the wish to become a nun as soon as her theses were published, if not before. But once again, her father pressured her to serve the family interests, and Agnesi acquiesced; she managed, however, to extract from him permission to abandon natural philosophy and focus exclusively on the study of mathematics – "the only province of the literary world where peace reigns," she believed. Pure mathematics allowed her to come as close as possible to meditating on the divine while pursuing a secular course of study; composing her famous *Instituzioni analitiche ad uso della gioventu' italiana* ("Analytical Institutions for the Use of Italian Youth") also served a religious function, furnishing her with the opportunity to bolster Catholicism through a textbook on the new analytical techniques, which conformed perfectly with the approved metaphysical and theological teachings of the Church. The book was published in 1748; four years later her father died, and Agnesi was free at last to abandon secular learning and devote herself to works of piety and charity.

The examples of Bassi and Agnesi appear to have inspired numerous fathers to promote "their daughters as great prodigies of learning in the hope of winning fame and fortune in the Republic of Letters." Not all of them, of course, proved to be prodigies of the first order, but the considerable resources and opportunities now made available to young women resulted in a minor educational revolution. Equally important for the present purposes, the examples of Laura Bassi and Maria Gaetana Agnesi also conferred newfangled respect on the mathematical sciences and natural philosophy – subjects that, though never replacing traditional humanistic disciplines, nevertheless became important to the formation of young women. In fact, as far as female prodigies were concerned, their measure was now taken, not by their ability to converse fluently in Latin or compose reams of poetry, but by their fluency in the exact sciences, the new criterion that separated the truly distinguished from the simply well turned out and polished.

The career of Cristina Roccati serves as a perfect example. Born in 1732, she received an education purposely calculated by an ambitious father to make her "another Laura." After rigorous private education in Rovigo, Roccati enrolled in 1747 as a student in Bologna; for three years she attended scientific lectures and experimental demonstrations at both the University of Bologna and the Bologna Institute for Sciences, as well as made the rounds as a welcome guest at various local salons. In 1751, she received her degree from the university, with none other than Laura Bassi accompanying her to the ceremony. From Bologna, Roccati proceeded to the University of Padua, purposely to perfect her understanding of Newtonian physics, before returning to Rovigo the following year. For the next quarter of a century, she lectured on physics and mathematics in Rovigo, helping to disseminate in the Veneto (much as Bassi had disseminated in Bologna) the legacy of the "immortal Newton" – whose theory of universal gravitation she praised as "one of the most necessary and most beautiful attributes that God has given to bodies."[20]

Singularity, however, could prove a double-edged sword. Defenders of reorientation in education paraded the examples of illustrious women, invariably generalizing from them the potential of all women. Such was the message of Sophia, the pseudonymous author of *Woman Not Inferior to Man* (1739), for whom the career of Laura Bassi was "a living proof that we are as capable, as any of the Men, of the highest eminencies in the sphere of learning, if we had justice done to us." Yet even Diderot, who willingly attributed the subordinate status of women to an archaic legal system and poor education, balked at the suggestion of the French philosopher Claude-Adrien Helvétius that women might be as capable as men of garnering the fruits of education. Helvétius' putative proof of women's equality, Diderot snickered, was the actual genius exhibited by a small number of women over the centuries; "And from this small number I am to conclude equal aptitude to genius in one and the other sex, and that one swallow makes a summer."[21]

Diderot's attitude is thus indicative of how little had changed over the course of a century. By the mid-eighteenth century, women had emerged as the audience of choice for publish-

INSTITUZIONI
ANALITICHE
AD USO
DELLA GIOVENTU' ITALIANA
DI D.NA MARIA GAETANA
AGNESI
MILANESE
Dell' Accademia delle Scienze di Bologna.
TOMO I.

IN MILANO, MDCCXLVIII.
NELLA REGIA-DUCAL CORTE.
CON LICENZA DE' SUPERIORI.

For Agnesi, composing her famous *Instituzioni analitiche* (Milan, 1748) served a religious function, providing an opportunity to bolster Catholicism through a textbook on the new analytical techniques, which conformed perfectly with the approved metaphysical and theological teachings of the Church. – NYPL-SIBL

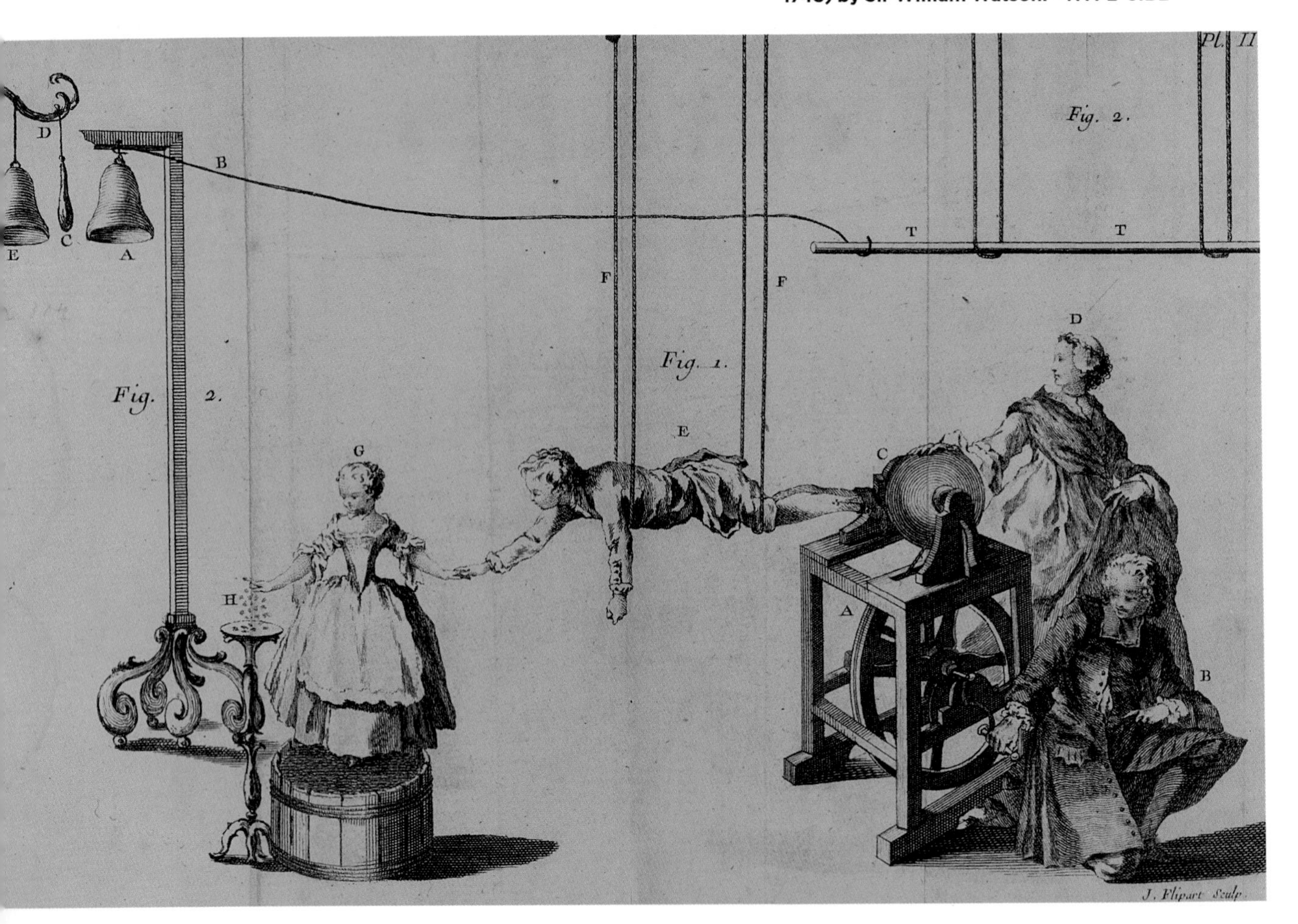

A popular eighteenth-century electrical experiment: a boy, suspended by silk cords, draws a charge from a generator; he transmits the charge to a girl standing on dried pitch, allowing her to attract and repel chaff. From a French edition of *Experiments and Observations Tending to Illustrate the Nature and Properties of Electricity* (Paris, 1748) by Sir William Watson. – NYPL–SIBL

ers of popular scientific and philosophical works; the salons of women became pivotal venues for the diffusion of new scientific ideas; and the political power wielded by women over (male) academic institutions was unprecedented. (Of Mme de Lambert, it was said that one had "to pass through her in order to get into the Académie française"; so, too, Lazzaro Spallanzani was convinced that it was not "a man's merit, but Venetian ladies," that determined "who was to become a lecturer at the University of Padua."[22]) Nevertheless, the active participation of women in the sciences remained marginal; traditional biases regarding the advantages of steering clear of esoteric subjects or, alternatively, of concealing whatever knowledge a woman had acquired remained strong. Mme de Lambert advised her daughter to "moderate" her taste for "extraordinary sciences," which she described as dangerous and conducive to pride. "Girls must maintain the same modesty in respect to the sciences as they do toward vice," Lambert admonished, cautioning her daughter to desist from "running after vain sciences" that were beyond her reach and not intended for women.[23]

Even more adamant on the subject was Lady Mary Wortley Montagu. Women are forbidden higher studies, she thundered, and are permitted only books that tend "to the weakening or effeminating of the mind." To "improve our reason or fancy [that] we have any" is considered "as in a degree criminal," and there is "hardly a creature in the world more despicable or more liable to universal ridicule than that of a learned woman." For all her railing, when it came to the education of her own granddaughter, she recommended that a special effort be made to "conceal whatever Learning she attains, with as much solicitude as she would hide crookedness or lameness. The parade of it can only serve to draw on her the envy, and consequently the most inveterate Hatred, of all he and she Fools." Montagu nonetheless endorsed her granddaughter's study of philosophy; though only "a few heads" were capable of undertaking Newton's calculations, she observed, "the result of them is not difficult to be understood by a moderate capacity."[24]

Montagu chose her words advisedly in lumping together "he and she Fools." Women's general censure of the superior minds of certain of their sisterhood could be as cruel as any misogynist's. Mme du Deffand's pen-portrait of Mme du Châtelet speaks for itself:

She was born with a fairly good mind; wishing to appear even cleverer she preferred the study of the most abstract sciences to more agreeable knowledge: she hopes in this peculiar way to attain to a greater reputation than all other women, and to a decided superiority over them.... However famous mme Du Ch[âtelet] might be she

In her *Institutions de physique* (1740), the celebrated Gabrielle Emilie Le Tonnelier de Breteuil, marquise du Châtelet (1706–1749), following upon her collaboration with Voltaire on his *Elémens*, made an original and impressive attempt to give Newtonian physics a metaphysical (and partly Leibnizian) framework. – Courtesy of Henri de Breteuil

> could not be satisfied if her praise were not sung, and this also she has achieved in becoming the acknowledged mistress of m. de Voltaire. It is he who gives her life its lustre, and it is to him that she will owe her immortality.[25]

Understandably, then, the celebrated du Châtelet wavered between modesty and assertiveness in her presentation of herself and her work. Her ambition was genuine; yet the

Mme du Châtelet wavered between modesty and assertiveness in her presentation of herself and her work. The *Institutions de physique* (1740) appeared without her name on the title page, but in its allegorical frontispiece, the muses of the various sciences watch admiringly as the marquise herself ascends toward the Temple of Nature, the gate of which is adorned with a portrait of Newton in the center, and portraits of Descartes and (perhaps) Malebranche to each side. – Burndy Library

Institutions de physique ("Institutions of Physics," 1740) appeared without her name on the title page, and the preface opened with a disclaimer that the book was the work of a mother who wished to instruct her son – a qualification adopted by more than one woman venturing into print. Nevertheless, the allegorical engraving she commissioned for this book – which made an original and impressive attempt to give Newtonian physics a metaphysical (and partly Leibnizian) framework –belies such modesty. In it, the muses of the various sciences watch admiringly as Mme du Châtelet in her own person ascends toward the Temple of Nature, the gate of which is adorned with a portrait of Newton in the center, and portraits of Descartes and (perhaps) Malebranche on each side. The dark, gray clouds she mounts, much like steps, will lead her to the Temple on high, irradiated in light, where the naked goddess of Truth gestures welcomingly. Yet this frontispiece must not be accepted at face value. Shortly after completing the book, the deeply ambivalent Marquise complained that God had refused her "any kind of genius"; she'd wasted her time "unraveling truths that others have discovered." But it did not take long for her deep opionion of her self-worth to come full circle again. Working on her translation of Newton's *Principia* into French, she wrote to Frederick the Great of Prussia: "it may be that there are metaphysicians and philosophers whose learning is greater than mine, although I have not met them."[26]

Du Châtelet's pride, alternating with resentment, frustration, and dissimulation, was no doubt shared by other women who sought to acquire more than a passing familiarity with abstruse learning. European women were generally dependent on the willingness of a parent or guardian to allow them an education, for whatever reason. Not surprisingly, many educated women lacked male siblings, and thus had received the scholarly attention most fathers would have lavished on a son. Suzanne Curchod, the object of the historian Edward Gibbon's early affection, for example, owed her status as a prodigy of sorts to the determination of her father – an obscure minister living "in the solitude of a sequestered village" near Geneva – to confer on his only daughter "a liberal and even learned education."

She "surpassed his hopes by her proficiency in the sciences and languages," and went on to become, not Gibbon's wife, but the celebrated Parisian salonnière Mme Necker. In a similar manner, the Viscount de Chateaubriand encountered while in exile in England during the 1790s the Reverend John Clement Ives, who loved to talk about Homer and Newton, and whose only daughter, Charlotte, "had become learned in order to please" him. And again, Maria Eimmart, the daughter of the Nuremberg astronomer Georg Christoph Eimmart, owed her scientific education to her status as an only child as well as to her extraordinary drawing skills: her numerous drawings of the phases of the moon proved invaluable to her father.[27]

Those women not fortunate enough to have ambitious (or understanding) fathers faced extraordinary obstacles. "From the moment I was first illuminated by the light of reason," recalled the celebrated Mexican writer Sor Juana Inés de la Cruz, "my inclination toward letters had been so vehement, so overpowering, that not even the admonitions of others – and I have suffered many – nor my own meditations ... have been sufficient to cause me to forswear this natural impulse that God placed in me." Determined to master the entire encyclopedia of learning, she proceeded to do so "having for a master no other than a mute book, and for a colleague, an insentient inkwell." The prodigy Sophie Germaine, who discovered mathematics at thirteen, encountered stiff resistance from her parents. According to one biographer, she overcame all obstacles "by getting up at night in a room so cold that the ink froze in its well, working enveloped with covers by the light of a lamp even when, in order to force her to rest, her parents had put out the fire and removed her clothes and a candle from the room."[28]

Overcoming prejudice concerning the propriety of women pursuing studies of a recondite nature was no easy matter, as Isabelle de Charrière (Belle de Zuylen) – Boswell's would-be spouse – discovered to her chagrin. She was refreshingly open about her unbounded ambition: "[my] vanity is boundless and boundless by gift of nature," she wrote. Though cognizant that fame is often "purchased at the expense of happiness," she was prepared to "submit to much for the sake of fame." The fame she eventually achieved came in the literary arena, but in her early twenties she had fixed her sights on the natural sciences as well. Writing to her secret correspondent Constant d'Hermenches

Isabelle de Charrière (1740–1805), known as Belle de Zuylen, was prepared to "submit to much for the sake of fame" and thumbed her nose at the specter of not finding a suitable mate: "If I were married," she wrote to a secret correspondent in 1764, "I would not give as many hours to the harpsichord or to mathematics, and that would distress me, for I want absolutely to understand Newton, and to become almost as good an accompanist as you." This portrait is by Maurice Quentin de La Tour. – © Musée d'art et d'histoire, Ville de Genève

Francesco Algarotti's *Il Newtonianismo per le dame* targeted the general reading public, and women in particular; the frontispiece to the Naples, 1737, edition is reproduced here. At the book's conclusion, his protagonist – a marquise – is informed that "the light of *Newtonianism* has dissipated the *Cartesian* Phantoms which deluded your Sight. You are now really a *Newtonian*." She remains, however, a passive recipient of diluted knowledge, advised explicitly not to aspire to more profound understanding. – Burndy Library

in early January 1764, Belle thumbed her nose at the specter of not finding a suitable mate: "If I were married, I would not give as many hours to the harpsichord or to mathematics, and that would distress me, for I want absolutely to understand Newton, and to become almost as good an accompanist as you." D'Hermenches was appalled, entreating his willful friend to eschew mathematics: "It shrinks the imagination, it desiccates the mind; these proofs are at the expense of feeling; one must believe, and validate, and taste without proofs." Belle was not dissuaded, at least not yet. She sought to allay d'Hermenches' fears, while boasting of the progress she had already made: "I am, however, much more advanced in it than you might think; I study all the properties of conic sections with the greatest diligence. My tutor, who is not polished and no flatterer, has told me that he has never seen a better aptitude, nor such rapid progress." She justified her passion for mathematics by citing its contribution to her well-being and to her ability to admire God's creation. "I would like to know of physics what is known in my time, and for that, one must have mathematics." To this end, she employed the mathematician Laurens Praalder as private tutor and enrolled at Utrecht in courses of experimental philosophy. For all her charm and bravado, however, her friends and family wore her down and she bade farewell to such abstruse studies.[29]

The list of obstacles faced by women in their pursuit of advanced education must also include condescension from the very individuals mobilized to their cause. Even royalty was not exempt from such condescension, as can be gathered from Queen Sophie Charlotte of Prussia; Leibniz, she complained in 1701, tended to treat everything with her very superficially, mistrusting her intelligence and rarely responding in detail to her queries.[30] Mme de Beaumer, the first female editor of *Le Journal des Dames*, found a similar attitude in Charles Joseph de Villers's *Journées physiques* ("Daily Journeys in Physics," 1761). In the wake of the return of Halley's Comet – another spectacular confirmation of Newtonian physics – de Villers sought to graft Fontenelle's nearly century-old model of sugar-coated, simplified astronomical dialogue onto contemporary (Newtonian) cosmology. As far as de Beaumer was concerned, such a literary piece was a sham. What women required was "serious instruction, not watered-down drivel," she fulminated, flaunting Mmes Dacier and du Châtelet as proof of the erudition of which women were capable. When the astronomer Joseph Jerôme de Lalande visited the twenty-one-year-old Sophie Germain in 1797, he suggested she read his popular *Astronomie des dames* ("Ladies' Astronomy," 1785). The indignant Germain responded that she had already read Laplace's more profound *Exposition du systéme du monde* ("System of the World"), published the previous year, and had no wish to read his primer. When Lalande rejoined that he did not think she could "understand the one without the other," the infuriated Germain all but threw him out of the house.[31]

Indeed, patronizing sentiments informed most treatises that sought to popularize the new science, specifically those directed toward women. Fontenelle's *Entretiens* initiates his heroine the Marquise into the mysteries of Cartesian cosmology, but the information it furnishes and the reasoning it inculcates do not allow for independent development beyond the boundaries set by her interlocutor. Fontenelle's objective is to keep the Marquise a permanent – albeit more informed – spectator. Nor was Algarotti's Marquise perceived any differently, except that at the conclusion of *Newtonianismo*, she is informed that "the light of *Newtonianism* has dissipated the *Cartesian* Phantoms which deluded your Sight. You are now really a *Newtonian*." For all intents and purposes, however, she remains a passive recipient of diluted knowledge, advised explicitly not to aspire to more profound understanding – for the "female sex prefers to feel rather than to know."[32]

One need not have been a Mme du Châtelet to strike an aggressive pose in parading scientific knowledge – and be derided for it. The resident English Consul to Florence wrote in 1745 to inform the writer Horace Walpole that his stepmother, then in town, had fallen into conversation with the commander of a Croat regiment, Baron von Andrási – "a very learned man for a hussar" – and, as was her habit, had turned the conversation to philosophical matters. When the baron failed to become "confounded on her mention of Newton," the Lady cried out: "perhaps, General, you have read Voltaire." No, retorted Andrási, "I have read Newton and understand him very well." Eleven years later, having heard that Lady Walpole propounded electricity-based theories to account for earthquakes, Walpole – whose dislike for the Lady ripened with age – retorted: "I had rather any dowager of my acquaintance should lump an earthquake under the chapter of miracles, than be forced to explain to her the natural process of it: I am sure she will not talk half so much nonsense upon it in a religious style, as she would in a philosophic one – I have known the time when I am sure you and I should have wished that my Lady Orford and my Lady Pomfret had studied St. Chrysostom instead of Sir Isaac Newton."[33]

We do not know the extent of Lady Walpole's understanding of Newton, but clearly it was not negligible. The attitude of her stepson, however, conjures up the perception of Bathsua Makin who, several decades earlier, quipped that to men, a learned lady resembled "a Comet, that bodes Mischief, when ever it appears."[34] Perhaps it was on precisely such grounds that another remarkable Newtonian, Mlle Ferrand, was determined not to make the world – beyond a small circle of friends – privy to her considerable intellect. Consequently, we know very little about her. She appears to have occupied for the French philosopher Etienne Bonnot de Condillac the same relationship that Mme du Châtelet occupied for Voltaire – a collaborator and a perceptive critic. In the preface to his *Traité des sensations* ("Treatise on the Sensations"), Condillac paid warm tribute to Ferrand, going so far as to claim that the treatise was "largely her work." Had Ferrand been alive when the book was published, such a tribute would

SCIENCE FOR CHILDREN

2a

1

In 1802, an English publisher defended the suitability of fairy tales for children, arguing that they would "amuse and improve children as much as an account of Sir Isaac Newton's Philosophy." John Newbery's immensely popular Tom Telescope [3; 4a–b] was undoubtedly on his mind, and his plea reflects the rapid reorientation of childhood education in the course of the eighteenth century, with an increasing advocacy of science as the kind of knowledge best suited to the needs and predilections of the young. Indeed, children, as well as women, had emerged as the audience of choice for publishers of popular scientific and philosophical works – Meil's *Spectacle* [1] is typical – and the subject matter was thought suitable even for royalty, as is evident from Baumeister's luxurious *The World in Pictures* [2a–b].

2b

3

Frontiſpeice.

Lecture on Matter & Motion.

THE
NEWTONIAN SYSTEM
OF
PHILOSOPHY

Adapted to the Capacities of young GENTLEMEN and LADIES, and familiarized and made entertaining by Objects with which they are intimately acquainted:

BEING

The Subſtance of SIX LECTURES read to the LILLIPUTIAN SOCIETY,

By TOM TELESCOPE, A.M.

And collected and methodized for the Benefit of the Youth of theſe Kingdoms,

By their old Friend Mr. NEWBERY, in St. Paul's Church Yard;

Who has alſo added Variety of Copper-Plate Cuts, to illuſtrate and confirm the Doctrines advanced.

O Lord, how manifold are thy Works! In Wiſdom haſt thou made them all, the Earth is full of thy Riches.

Young Men and Maidens, Old Men and Children, praiſe the Lord. PSALMS.

L O N D O N,

Printed for J. NEWBERY, at the BIBLE and SUN, in *St. Paul's Church Yard.* 1761.

4a

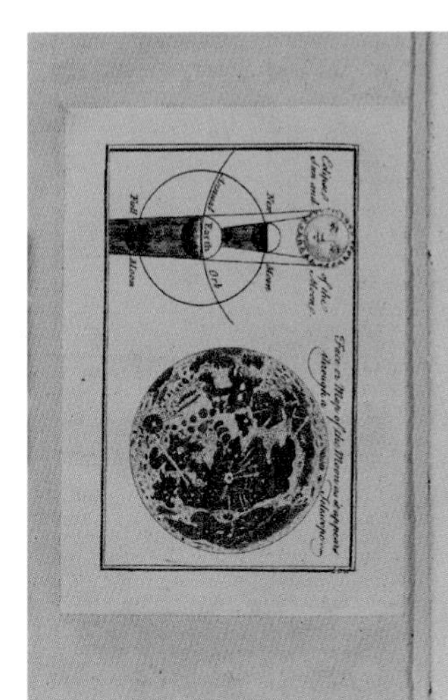

VAN HET ZONNESTELZEL. 29

die de Maan verbeelt, een weinig naar een van beide zyden, en met het oog geplaatst als te vooren, zult gy een gedeelte van den Oranje-appel gewaar worden, welke nu in zulken ſtand zal zyn, dat eene regte lyn van een gedeelte van den Oranje-appel of Zon getrokken kan worden tot een gedeelte van den Tol of Maan, zonder den Bal te raaken, die de Aarde vertoont; en in dezen ſtand zou de Maan gedeeltelyk verligt, en de Verduiſteringe alleen *gedeeltelyk* zyn.

Eene *Zonneverduiſteringe* wordt veroorzaakt door den ſtand der Maane tusſchen de Zon en de Aarde, welke het ligt van de Zon belet tot dat gedeelte van de Aarde te koomen, het geen wy bewoonen . . . . . . Wanneer de Maan het ganſche Lighaam van de Zon voor ons verbergt, is het eene *geheele* Verduiſteringe; maar indien het niet geheel is verborgen, is het eene *gedeeltelyke*. Eene Verduiſteringe van de Zon valt nooit voor dan op *nieuwe* Maan; nog eene Verduiſteringe der Maane dan wanneerze *vol* is.

De Maan, betreffende derzelver Stoffe en Gedaante, ſchynt niet zeer ongelyk aan onze Aarde, gelyk gy door deze Kaart kunt merken. De heldere of ſchitterende gedeelten worden onderſteld hooge verligte ſtreeken Lands te zyn, als *Bergen*, *Eilanden &c.* En de donkere gedeelten, verbeeld men zig, dat *Zéen*, *Meiren*, en *Valeijen* zyn, welke maar weinig ligt wederomkaatzen. Dog hier van is geene zekerheid.

De Aarde meet, door haare omwentelinge om

4b

HEEL-AL enz.

s onze Aarde bewoond
y op de Aarde onbekwaam
kken, wat zoort van In-
Wat my betreft, ik ver-
dezen onbepaalden afgrond.
e, dat de Zon, welke aan
leeven geeft, alleen eene
Heerlykheid; en de Lugt,
onderhoudt, is, als het
nzer Neusgaten.
nderſteun my, terwyl ik
uwe wonderbaare Voort-
ouwe; nadien het geen yde-
yke nieuwsgierigheid zy,
t Onderzoek leidt; maar
geerte, om alleen de uiter-
e Glorie te bemerken, op
en uwe Barmhartigheid aan
het Menschdom moge verheerlyken.

VAN HET ZONNESTELZEL.

Ons *Zonneſtelzel* (*Solar Syſtem*) bevat de Zon in het middelpunt, en de Dwaal- en Staartſterren, rondſom dezelve beweegende . . . . Eylieve ziet op de Figuur aan de andere zyde, alwaar ik de Zon en de Dwaal-ſterren, in hunne verſcheidene Loopkringen met hunne byzondere afſtanden van de Zon, en van elkanderen, hebbe afgebeeld; te gelyk met den Loopkring van eene Comeet of Staartſterre.

De Dwaalſterren, gelyk ik alrede hebbe aangemerkt, zyn Lighaamen, die gelyk Sterren ſchynen, maar zyn niet ligtende of helder

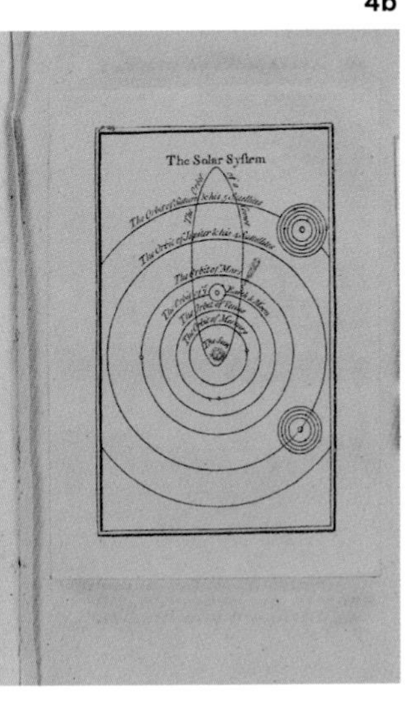

1: Johann Wilhelm Meil, *Spectaculum naturae & atrium* (Berlin, 1761–65). – Harvard College Library

2a–b: Joseph Anton Ignaz Baumeister, *Die Welt in Bildern* (Vienna, 1788–94). – Bildarchiv d. ÖNB, Vienna

3 (shown actual size): Tom Telescope [John Newbery], *The Newtonian System of Philosophy, Adapted to the Capacities of Young Gentlemen and Ladies ... Being the Substance of Six Lectures Read to the Lilliputian Society* (London, 1761). – NYPL–Berg Collection

4a–b: Tom Telescope [John Newbery], *Philosophie der tollen en ballen* (Middelburg, 1768). – NYPL–Rare Books Division

Mlle Ferrand was another remarkable Newtonian. The French philosopher Etienne Bonnot de Condillac, for whom she was a collaborator and a perceptive critic, paid warm tribute to her, even claiming that his *Traité des sensations* was "largely her work." In this portrait, "Mademoiselle Ferrand Meditating on Newton," painted by Maurice Quentin de La Tour in 1752, she is depicted alongside a large folio volume of Newton. – Reproduced courtesy of the Collection of Bayerische Hypo- und Vereinsbank AG in the Alte Pinakothek, Munich / © Joachim Blauel – ARTOTHEK

undoubtedly have been omitted. In sharp contrast to Condillac, a rancorous letter writer describes Ferrand as "a person of little wit and a disagreeable manner, but who knew geometry." Such imputed disagreeableness, however, is belied by the exclusive Parisian philosophical salon she hosted, atypically financed by the celebrated guests that frequented it. Perhaps the best glimpse of her character is offered by the portrait "Mademoiselle Ferrand Meditating on Newton," painted by Maurice Quentin de La Tour. Seated at her desk, in untidy morning clothes, with a large folio volume by Newton open on a stand, her mien suggests both comely beauty and bemused irritation, as if she has been inappropriately interrupted in her private room, at an unsocial hour.

The spectacle of women in avid pursuit of mathematics and natural philosophy, then, elicited some begrudging respect, but more commonly biting satire. The publication of Molière's *Les femmes savantes* ("The Learned Ladies") in 1673 set the tone. Early on in the play, a young woman is chastised by her elder sister for conceiving of bliss in terms of "an idolized husband and a bunch of brats"; she is urged instead to cultivate her mind – as their mother does – and wed herself to philosophy. The men rise up against precisely such a notion. The younger sister's suitor declares that "learned women are not to [his] taste"; the father bemoans the passing of those times when women knew their place: women today wish only to become authors and "no science is too deep for them." As it stands now, his household is in disarray, his food neglected, for even the servants have been trained to assist the women to research the loftiest secrets of astronomy.[35]

Although Molière's satire excoriated the excesses of overzealous "femmes savantes," what resonated with contemporary audiences was the broad conclusion that, as one of Molière's characters puts it, "it is not seemly, and for many reasons, that a woman should study and know many things." Molière thus contributed the most powerful tonic to date to the public display of scientific knowledge by women. Indeed, fear of ridicule became a central concern of women curious about science, so much so that authors of popular books intended for them, such as Benjamin Martin's *Young Gentleman and Lady's Philosophy*, were forced to address the issue and assuage concerns.

But it was all to no avail. Mirth and ridicule proliferated in the literature of the day. In one of the early issues of the *Spectator*, Joseph Addison enumerated the contents of a sizable library of the "learned" Leonora, most of which she purportedly acquired not for use, but "either because she had heard [the books] praised, or because she had seen the Authors of them." The books for ostentation included a complete run of Newton's works.[36] Joseph de Maistre uses flippancy as a way to describe a beautiful young woman's eagerness to inquire what prevents her from knowing as much astronomy as Newton. Nothing at all, de Maistre would have answered if asked; the stars themselves, honored to be ogled by her, would rush to reveal their secrets. Cleverer still was the poet and military general François-Joseph de Beaupoil de Sainte-Aulaire. Having attended the salon of the Duchesse du Maine – who "believed in herself as she believed in God or Descartes, without examination and without discussion" – Sainte-Aulaire found himself chided by his hostess for failing to participate in the discussion on Descartes and Newton she had initiated. Sainte-Aulaire instantly produced an impromptu play on words vis-à-vis Descartes to avenge himself:

Bergère, détachons-nous
De Newton, de Descartes:
Ces deux espèces de fous
N'ont jamais vu le dessous
Des cartes, des cartes, des cartes!

(Shepherdess, let's detach
From Newton, from Descartes:
These two species of fools
Have never seen the backs
Of cards, of cards, of cards.)[37]

A similar scoffing tone about female pursuit of science informs Colnet du Ravel's *L'art de diner en ville*:

She resolved with one word,
while knotting her ribbon,
These great questions that
stupefied Lagrange.
One sees on her dressing table
an Euler, a Pascal,
Dirtied and smeared with
vegetable red [dye].
She finds in Newton some
I know not what appeal
And algebra has an
inexpressible charm for her.
The evening, in a dungeon,
with a curious glance,
Questioning the skies,

at the end of a lens,
Her observant eye pursues
the comet there:
Lalande every year
steals a planet for her.[38]

"A Female Philosopher in Extasy Solving a Problem," by Maurice Quentin de La Tour, 1759. – Colonial Williamsburg Foundation

The pictorial medium was the most effective at satirizing the pretensions of women to learning beyond their capabilities – and the consequences of such pursuit to their "virtue." La Tour's "A Female Philosopher in Extasy Solving a Problem," for example, conjures up the hilarious image of a young woman in a state of (sexual) rapture, having just grasped, it seems, the force of a demonstration in Euclid's *Elements* (a copy is on the table, along with a globe). Certainly, the befuddled presence of a male servant, who is peeping in to discover the nature of the commotion, underscores the aberrant nature of her behavior. Robert Sayer's 1793 print "Viewing the Transit of Venus" offers an only slightly less ribald tribute to female devotion to science. Here, an elegant lady observes the planet Venus through a telescope – the title alludes to the celebrated transit of the planet across the sun that occurred in 1761 and 1769 – while her mentor takes advantage of her intent gaze and forward-leaning position to use his own looking-glass, a monocle, to gaze down at her cleavage. A grinning satyr in the garden provides a commentary on the pursuits of Venus, both planet and goddess of love.

Even more explicit is John Lodge's print "A Philosopher Giving Lectures on the Use of the Globes." Once again, a play on words captures the diametrically opposed interests of the woman seeking enlightenment and the men in a position to gratify her wishes. While the innocent maiden clutches a copy of Aristotle, her tutor exclaims (as he removes a piece of clothing): "Behold my Pupil, these Celestial Globes. Behold the milky way how heavenly bright." A character standing by ogles the lady's breasts with a telescope, crying out, "I could wish to observe the Planet Venus with the naked — Eye. How wonderful are the works of Nature." A third is "Planet struck," while a fourth is speechless, his hand busy at his crotch. Here, a goat by the woman's side stands in for the satyr as a silent pundit.

"Viewing the Transit of Venus," by Robert Sayer, 1793. – Courtesy of the Lewis Walpole Library, Yale University

"A Philosopher Giving Lectures on the Use of the Globes," by John Lodge. – Courtesy of the Lewis Walpole Library, Yale University

6

In the preliminary discourse to his *Histoire naturelle, générale et particulière*, Georges Louis Leclerc, comte de Buffon, offered a naturalistic account of the formation of the solar system: a comet chipped off part of the sun's mass; the molten matter eventually condensed into the several planets and their satellites. This allegorical engraving from Volume 1 of the first edition (Paris, 1749) depicts the act of creation.
– NYPL–General Research Division

ALL WAS LIGHT

Edmond Halley's contribution to the first edition of the *Principia* included a Latin ode celebrating Newton's supreme success in discovering the inviolable laws of the universe, laws that even "the creator of all things, while he was setting the beginning of the world, would not violate." Because of Newton, Halley wrote, "we now have the secret keys to unlock the obscure truth; and we know the immovable order of the world, And the things that were concealed from the generations of the past."[1] Within half a century, there reigned a perception that the revolution effected by Newton and his successors had ushered in a brave new age of reason and light that went well beyond the natural sciences to include the totality of human knowledge. Alexander Pope's celebrated couplet gives voice to the perception of Newton as God's emissary in the discovery of the laws of nature:

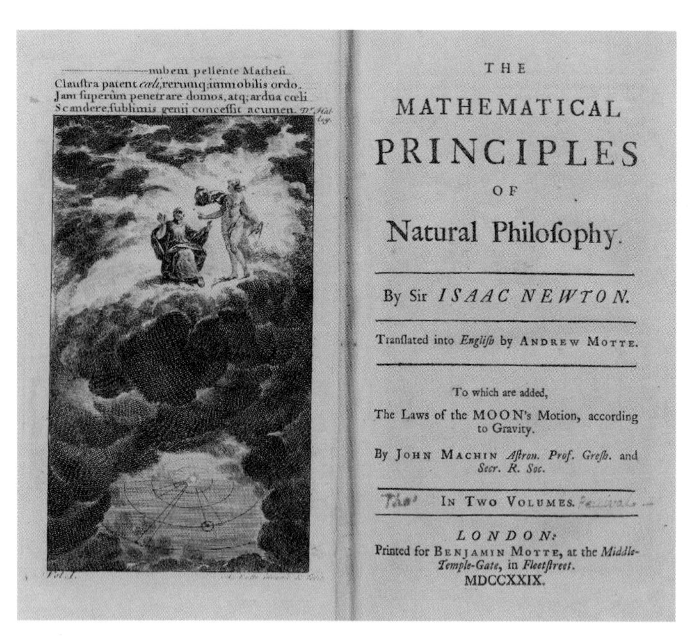

——nubem pellente Matheſi
Clauſtra patent *cœli*, rerumq; immobilis ordo.
Jam ſuperùm penetrare domos, atq; ardua cœli
Scandere ſublimis genii conceſſit acumen.

THE
MATHEMATICAL
PRINCIPLES
OF
Natural Philoſophy.

By Sir *ISAAC NEWTON.*

Tranſlated into *Engliſh* by ANDREW MOTTE.

To which are added,
The Laws of the MOON's Motion, according to Gravity.

By JOHN MACHIN *Aſtron. Prof. Greſh.* and *Secr. R. Soc.*

IN TWO VOLUMES.

LONDON:
Printed for BENJAMIN MOTTE, at the *Middle-Temple-Gate*, in *Fleetſtreet*.
MDCCXXIX.

The frontispiece to Andrew Motte's English translation of the *Principia* (London, 1729) displays a sage Newton seated in some sort of heavenly abode, with Naked Truth, a compass in hand, by his side; beneath him, dark clouds roll back to reveal a diagram of the solar system he had explained. - NYPL- Rare Books Division

Nature, and Nature's Laws lay hid at Night.
God said, *Let Newton be*! and All was
Light.[2]

Divine sanction for Newton's bringing forth light from darkness propels the metaphor from the literal stage – Newton's radical theory of light and colors – to the primal stage of creation set forth in the opening verses of Genesis:

In the beginning God created the heaven and the earth. And the earth was without form, and void; and darkness was upon the face of the deep. And the Spirit of God moved upon the face of the waters. And God said, Let there be light: and there was light.

The twin images of light and, by extension, reason became central to Enlightenment ideology. "There is a mighty light which spreads itself over the world," gloated the Earl of Shaftesbury in 1706, "especially in these two free nations of England and Holland, on whom the affairs of all Europe now turn." D'Alembert used the term "light" no fewer than nineteen times in his "Preliminary Discourse" to the *Encyclopédie*, while his co-editor on the project, Denis Diderot, taking stock of their endeavors in 1760, observed that "philosophy marches forward with giant steps, and light accompanies and follows it." An error, conversely, "throws shadows on surrounding truths." Diderot was equally adamant about the source from whence such reason came: "Without the English, reason and philosophy would still be in the most despicable infancy in France."[3]

The motifs of light and reason found visual representation almost immediately. The 1729 English translation of the *Principia* displays a sage Newton seated in some sort of heavenly abode, with Naked Truth, a compass in hand, by his side; beneath him, dark clouds roll back to reveal a diagram of the solar system he had discovered. An inscription, lifted from Halley's ode, provides literal accompaniment:

mathematics drives away the cloud.
Error and doubt no longer encumber us
with mist;
For the keenness of a sublime intelligence
has made it possible for us to enter
The dwellings of the gods above and to
climb the heights of heaven.[4]

The most famous representation of the light motif is embedded in the frontispiece of that Enlightenment monument to reason, the *Encyclopédie*, published in seventeen folio volumes of text and eleven volumes of plates between 1751 and 1772. The spectacular frontispiece, painted by Charles-Nicholas Cochin in 1764, was engraved and distributed to subscribers in 1772 with the following "explication":

Beneath a temple of Ionic architecture, sanctuary of Truth, we see Truth wrapped in a veil, radiant with a light which parts the clouds and disperses them. On the right of Truth, Reason and Philosophy are engaged, the one in lifting the veil from Truth, the other in pulling it away. At her feet, kneeling Theology receives her light from on high. Following the line of figures, we see grouped on the same side Memory, and Ancient and Modern History; History is writing the annals, and Time serves as a support for her. Grouped below are Geometry, Astronomy, and Physics. The figures below this group represent Optics, Botany, Chemistry, and Agriculture. At the bottom are several Arts and Professions that originate from the Sciences. On left of Truth we see Imagination, who is preparing to adorn and crown Truth. Beneath Imagination, the artist has placed the different genres of Poetry: Epic, Dramatic, Satiric, and Pastoral. Next come the other Arts of Imitation: Music, Painting, Sculpture, and Architecture.[5]

The radiant light dispelling the clouds of ignorance emanates, not from heaven, as had been the case in traditional iconography, but from the Temple of Truth. In fact, Theology actually turns her back on Truth, supplicating on her knees and seeking illumination from above – in vain, it seems, for no other light is forthcoming. Reason, in contrast, wearing the crown traditionally reserved for Theology, is assisted by Philosophy in removing the veil shrouding Truth. Philosophy also extends her left hand as if to restrain Theology. Even Imagination is preparing to adorn Truth with a bouquet of flowers. The iconoclastic representation struck Diderot the first time he set eyes on Cochin's design: "The philosophers have their eyes fastened on Truth; proud Metaphysics tries to divine her presence

Upon first seeing Charles-Nicholas Cochin's frontispiece to the *Encyclopédie* (designed 1764; engraved 1772), Diderot was struck by its iconoclasm: "The philosophers have their eyes fastened on Truth; proud Metaphysics tries to divine her presence rather than see her. Theology turns her back and waits for light from on high." – Private Collection

This frontispiece to a 1743 reprint of a seventeenth-century treatise on glassmaking visualizes the new Enlightenment conviction that even imagination must be ruled by reason, with, on the left, Reason and Experience illuminating the realm of Truth and, on the right, the Realm of Darkness, where Imagination dwells. – The Danish Pharmaceutical Library, Copenhagen

rather than see her. Theology turns her back and waits for light from on high."[6]

The subjugation of imagination to reason also became a potent Enlightenment motif. In a poem addressed to a fellow student, the great Swiss Newtonian physiologist Albrecht von Haller cautioned the mathematician Johann Gesner to take heed now that he was "about to follow in Newton's footsteps and thus enter nature's secret councils, led by the Newtonian art of measurement, the infallible bridle of the imagination."[7] Haller's sentiments found corroboration in the 1765 edition of the most popular iconography sourcebook, Cesare Ripa's *Iconologia*: "Imagination has been converted to faith in Reason, and the light of reason will liberate her from her age-old trauma by rendering those chimeras superfluous." The frontispiece of a 1743 reprint of *Ars vitraria experimentalis* ("The Experimental Art of Glassmaking") translates the idea into a striking image: "On the left, Reason and Experience illuminate the realm of Truth, where *lux veritatis* reigns. Opposed to this on the right one sees the Realm of Darkness harboring Imagination, who blows empty breasts of *amentia* as a grotesque ornament into the lap of Understanding, inscribing them with the names of all those who err, producing phantoms. A troublesome youth pinches her foot, while the background figure of a donkey-eared Pegasus testifies to the blind faith and impudence – *temeritas* – which Imagination obeys. Reason's enlightened ascent in opposition to Imagination is openly proclaimed here."[8]

A specifically Newtonian spin on the complex relations between reason and imagination was forged by Louis-Sébastien Mercier, whom we encountered in an earlier chapter as a eulogist for Descartes. Subsequently, Mercier had quite a distinguished career as a man of letters, and many of his writings attest to his growing admiration of Newton. Of particular interest is a little-known piece, entitled "Sommeil" ("Sleep"), published in a collection of short essays and reflections, *Mon bonnet de nuit* ("My Night Cap"). It deserves to be quoted at some length:

> Newton sleeps! In an instant, that active and penetrating quality which gave life to the most abstruse sciences, which unravelled the system of the universe with so much clearness and precision, falls into darkness and confusion, and no longer forms any other than a heap of confused and erroneous ideas. Instead of those firm and fertile principles, it follows fleeting phantoms, and it is given up to ridiculous perceptions. The mind of the man of genius, who pursued truth with such astonishing sagacity, is abandoned to the most inordinate irregularity. Grotesque figures replace the most sublime geometrical lines; there is no longer any harmony in that head which astonished his fellow creatures.... But a ray of the sun opens Newton's eyes; he awakes, and instantly resumes his vigorous faculties; they rally like disciplined soldiers, who, at the first beat of the drum, are no longer scattered, but form one body.[9]

The significance of Mercier's piece goes beyond its adulatory stance toward Newton and his genius. Quite likely, it inspired Francisco Goya's "El sueño de la razon produce monstruos" ("The Sleep of Reason Produces Monsters"). A copy of the French edition of Mercier's work was owned by a close friend of Goya's, and the latter may have borrowed it. It is even more likely that Goya owned, or read, the collection in its Spanish translation, which, unlike the French original, actually opens with "Sleep."[10] Goya's etching was intended as the frontispiece to a series of some eighty prints, the *Caprichos* ("Whims"), but ended up as plate number 43 of the series, first published in 1799. Here, the artist is seen hunched over his desk, fast asleep. Various creatures of the night surround him, obviously alarming the sharp-eyed lynx by his side. An inscription accompanying the second version of the print clarifies Goya's intent: "The author dreaming. His only purpose is to banish harmful ideas commonly believed, and with this work of *Caprichos* to perpetuate the solid testimony of truth." To further tie Goya to the Enlightenment tradition, the light emanates from Goya, not from any

"The Sleep of Reason Produces Monsters," one of the most famous etchings from Goya's *Caprichos* (1799), was almost certainly inspired by Louis-Sébastien Mercier's short essay "Sleep," which contemplated whether the mind of a genius like Newton, "who pursued truth with such astonishing sagacity," suffered darkness and confusion while he slept. - NYPL-Print Collection

In this 1832 allegorical sketch, a glorification of science by the British geologist and amateur artist Henry de la Beche, light from a gas lantern has replaced the traditional torch as the symbol of Light dispelling the clouds of Ignorance and Superstition. – Courtesy of the Department of Geology, National Museum of Wales

divine source, suggesting that the mind has the power to repel any monsters that may gather around. The published version was captioned: "The Sleep of Reason produces monsters." A contemporary explanation, either by Goya or by a friend, interprets the imagery as a warning: "Imagination forsaken by Reason begets impossible monsters."[11]

Scholars disagree about the extent to which Goya embraced the Enlightenment faith in reason. Though his faith was probably not as unbounded as Diderot's, in this period of his life Goya appears to have accepted reason's supremacy. His imagery certainly offers a striking parallel to his putative source of inspiration, Mercier, and the monsters the latter had conjured up for Newton's sleep. By 1799, however, the period of romanticism and revolution was in full sway, tamping down nearly a century of glorification of reason. But by the third and fourth decades of the nineteenth century, however, romanticism had lost its allure, and confidence in the power of reason and science soared with a vengeance. The British geologist, and amateur artist, Henry de la Beche exemplifies this renewed confidence in his "The Light of Science Dispelling the Darkness Which Covered the World" (1832). Despite the nineteenth-century paraphernalia, the imagery hearkens back to Enlightenment motifs: a confident woman, fashionably attired, marches resolutely toward a globe, a geologist's hammer in her right hand, a gas lantern in her left. The beam of light she projects forces the dark clouds engulfing the globe to retreat, as the creatures of the night seek cover.[12]

This nineteenth-century celebration of the power of science bore out the confidence expressed by Condorcet exactly fifty years earlier, in his 1782 discourse before the Académie Française: "Ignorance and error still breathe, it is true: but these monsters, the most formidable enemies of man's happiness, drag with them the mortal arrow that strikes them; and their very shrieks that terrifies you, do nothing but prove how much the blows which they've received were firm and terrible."[13] Obviously, Newton was not single-handedly responsible for this apparent triumph of science; but in the minds of most of his contemporaries, it was his supreme accomplishment, in and of itself, that established the natural sciences on new and more secure foundations and raised them to unprecedented heights. By becoming science personified – notwithstanding the contribution of many others to the elucidation and growth of Newton's ideas – Newtonian science also became the model to emulate, the manifestation of "superior knowledge" that summoned all other learning to reorient itself along similar lines.

To Be a Newton

Isaac Newton was the greatest and the luckiest of mortals: the greatest because he discovered the law of universal gravitation, the luckiest because there exists but one universe. This tribute by the celebrated eighteenth-century mathematician Joseph-Louis Lagrange encapsulates the singular position that the author of the *Principia* and the *Opticks* came to occupy as the acme of human potential. More than mere hyperbole, this possibility fired the imagination and ambition of generations to come.

"Today I am twenty-one years old," the physicist Joseph Fourier noted in a postscript to a letter on March 22, 1789; "by this age Newton and Pascal had already earned their claim to immortality."[14] Fourier's *cri de coeur* was widely shared, not least by Napoleon, who regaled French savants with remarks regarding his youthful ambition to transcend Newton. Had he not been called to serve his nation, Napoleon told the poet Népomucène-Louis Lemercier, he would have embarked on the exact sciences and followed the path of Galileo and Newton. "And since he had constantly succeeded in his great enterprises," boasted the emperor, "truly I should have been equally distinguished by my scientific labours. I should have left behind me the remembrance of great discoveries. No other kind of glory would have tempted my ambition." On another occasion, when Napoleon shared with the mathematician Gaspard Monge his youthful ambition, the latter cited Lagrange to suggest that the emperor had come on the scene too late. Not at all, Napoleon replied; Newton merely "solved the problem of planetary movement," whereas what he "hoped to do was to discover how movement itself is transmitted, through infinitesimal bodies."[15]

As inspiring as Newton's achievement could be, it could also inhibit. The Scottish philosopher Henry Home, Lord Kames, made precisely this point in 1774: "the great Newton, having surpassed all the ancients, [had] not left the moderns even the faintest hope of succeeding him: and what man will enter the lists who despairs of victory?"[16] Kames went so far as to attribute the decline of mathematics in England to the inability of Newton's followers to measure up to the master. Across the channel, two decades earlier, Diderot had predicted the impending demise of mathematics as the long-term consequence of the new "Pillars of Hercules" erected by Newton's followers – an edifice so forbidding that "no-one [would] pass beyond."[17]

Ironically, the best proofs of the daunting shadow cast by Isaac Newton for almost two centuries were the intermittent, but persistent, attempts to elude it. The brash twenty-one-year-old Christian Friedrich Westfeld articulated precisely such an impulse and the reasoning behind it. His 1767 *Die Erzeugung der Farben* ("The Production of Colors") depicted the Newtonian heritage as an insufferable authority in need of shaking up if the scope of optical investigation was to be revitalized. The theory he proposed, Westfeld claimed, was most likely fanciful; indeed, he scarcely believed in it himself. Yet he was emboldened to publish it "in order to bring back into vogue the investigations of scholars into colors, or, indeed, into the nature of light," for, following Newton, "all efforts in this part

GLORY AND GRAVITY

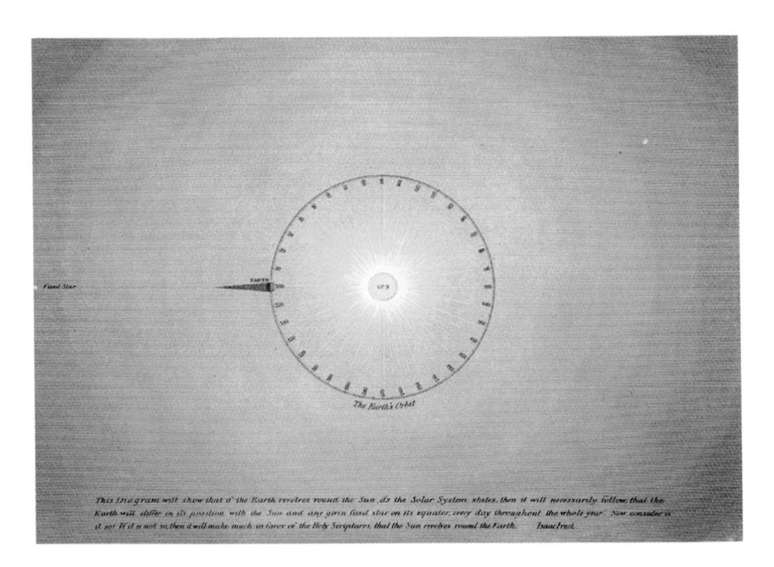

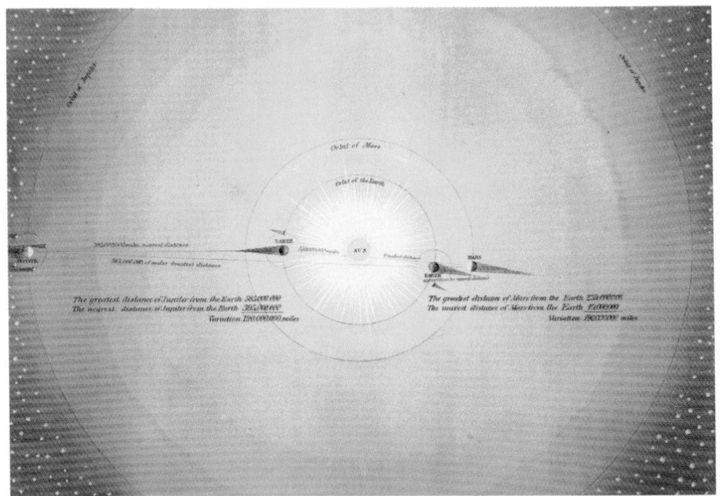

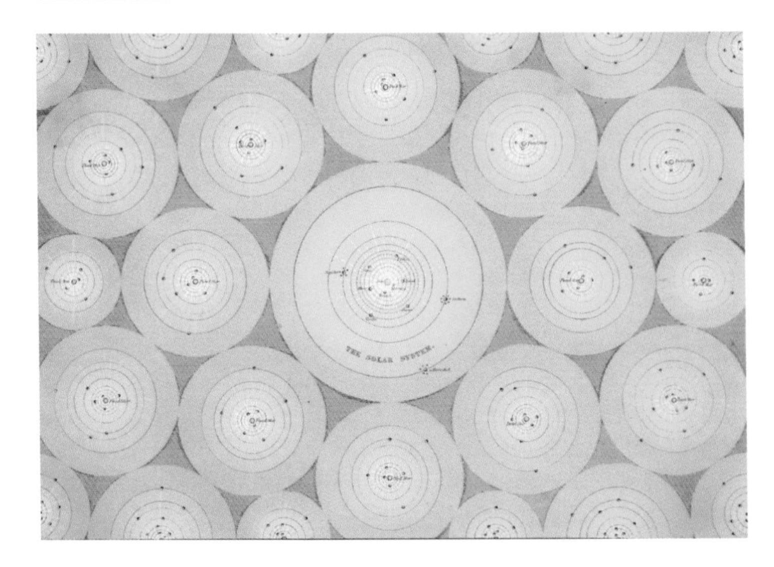

In 1846, Isaac Frost published his *Two Systems of Astronomy*, an illustrated comparison (intended for the edification of the Muggletonians, a minor British religious sect) between the Newtonian system of the world (represented on this page) and the one based on the Holy Scriptures (represented on the facing page). The key difference between the two systems, the author noted, is that in the system according to Scripture, the earth – not the sun – is fixed in the center, and is far larger in bulk and in size than the sun. The moon is a luminous body that gives light to the earth, as do the stars of the firmament. Nor is it possible to "behold the infinite space" when looking from the surface of the globe, as the Newtonians claimed, "because the firmament prevents us from doing so." – Courtesy of Adler Planetarium & Astronomy Museum, Chicago, Illinois

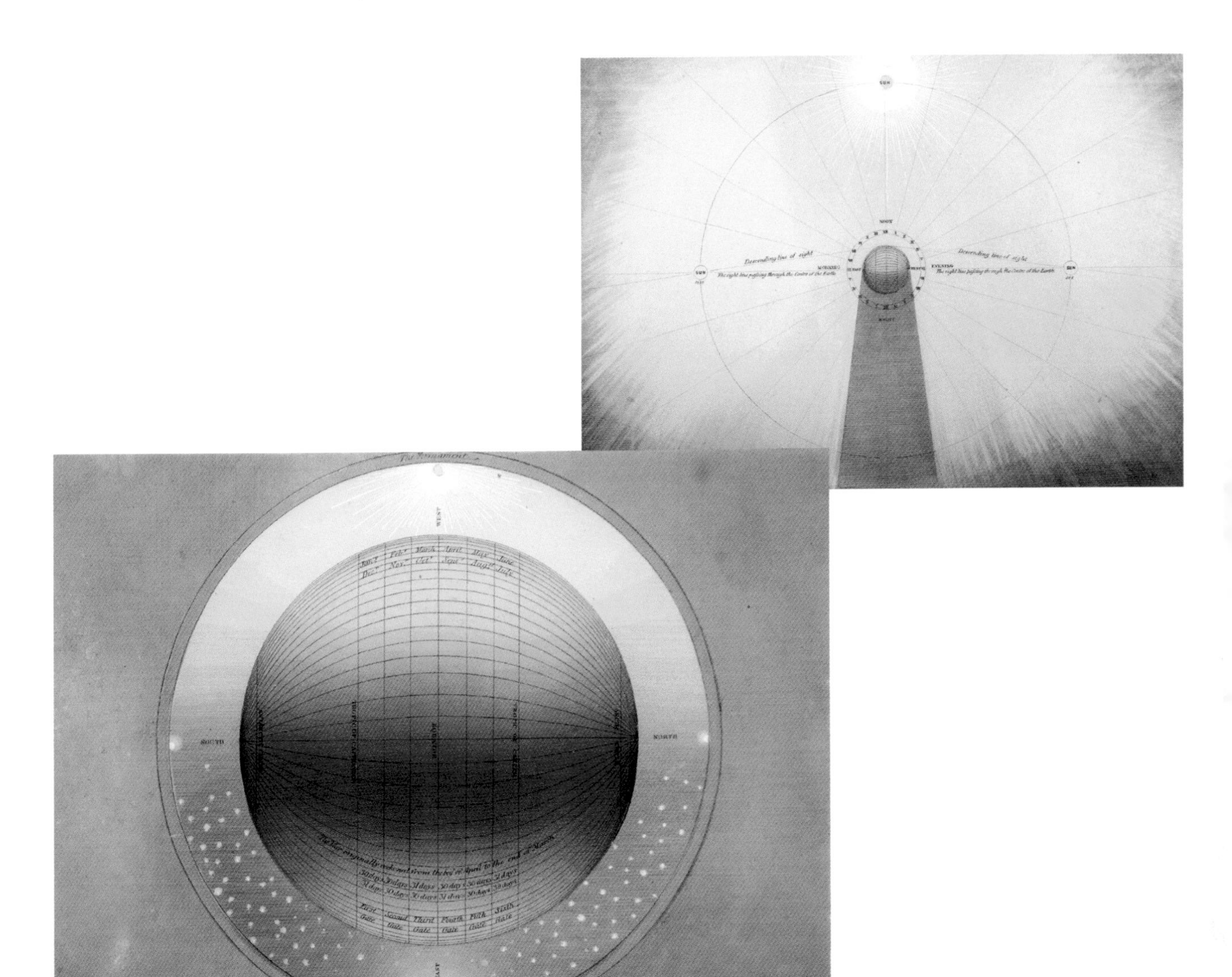
Descending line of sight
The right line passing through the Centre of the Earth
EVENING
Descending line of sight
The right line passing through the Centre of the Earth
WEST
SOUTH
NORTH
EAST
First Gate
Second Gate
Third Gate
Fourth Gate
Fifth Gate
Sixth Gate

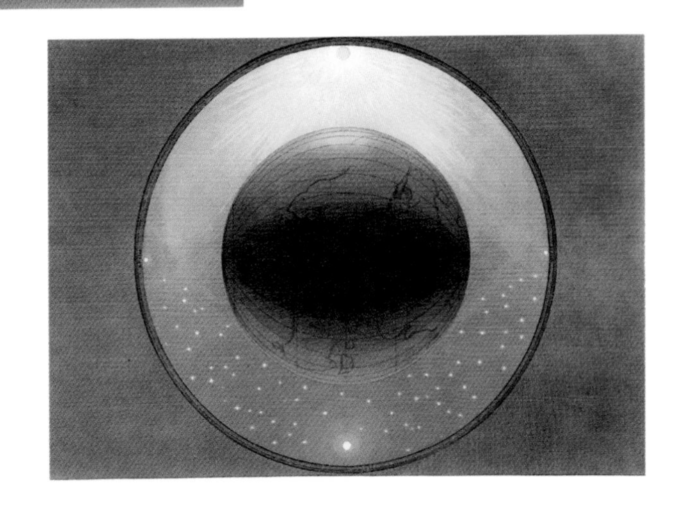

of natural philosophy have stopped." "It is really not good," Westfeld observed, "when too great a man treats a physical matter."[18] A far more talented man of science was Alexis-Claude Clairaut, who discovered a discrepancy between Newton's claims regarding the moon's rotation and the observed results, and so abandoned universal gravitation altogether on the grounds that "a theory which gave a value very divergent from the truth, ought to be 'condemned without appeal.'" Discovering his own error in calculation, Clairaut retracted in 1749 and obliquely "atoned" for his former rashness by imputing to the opaqueness of the *Principia* the penchant to "destroy" Newton's system, "without examining any of the calculations on which it had been based," or by presuming that metaphysics sufficed to prove the impossibility of attraction.[19]

Not surprisingly, those practicing in domains considered less rigorous than rational mechanics – or those aware that their scientific claims rested on somewhat shaky and controversial foundations – were most prone to battling the Newtonian legacy. The British natural philosopher and theologian Joseph Priestley, for one, employed the metaphor of a town to suggest the unfair lot of those who, like himself, sought to make contributions to disciplines already revolutionized by the likes of Newton. The construction of the first few streets, he noted, "makes a great figure, is much talked of, and known to every body; whereas the addition of, perhaps, twice as much building, after it has been swelled to a considerable size, is not so much as taken notice of." For Priestley, the exciting new field of electricity offered a new town, so to speak, more amenable to newcomers because no particular preparatory knowledge was required: "New worlds may open to our view, and the glory of the great Sir Isaac Newton himself, and all his contemporaries, be eclipsed, by a new set of philosophers, in quite a new field of speculation."[20]

In contrast to Priestley's determined, albeit respectful, attempt to demythologize Newton, Jean-Paul Marat chose to battle the Newtonian legacy largely by self-aggrandizement. Having failed to obtain a favorable reception from the Académie des Sciences for his theories regarding heat, light, and electricity – theories at once heavily dependent on Newtonian concepts and experiments, and hostile to Newtonian conclusions – the future principal player in the French Revolution took his case to the popular press. There, he and his friends launched a campaign to excoriate the scientific establishment and raise his standing above Newton's. Marat "flies from discovery to discovery with an astonishing rapidity," the *Courier de l'Europe* informed its readers in October 1781. His optical discoveries awed those physicists who remained faithful "with a too blind faith to the prophecies of Newton." Marat challenged "this idol of the day," defying "those who are most committed to his cult." The likely author of this article was none other than Marat himself, who the same year also penned an article that appeared in the *Journal de Littérature, des Sciences et des Arts*: "The revolution that M. Marat has just made in optics has created such a sensation among the Physicists who cultivate this science that they still have not recovered from their astonishment. Those who actually are most inclined to accept innovations cannot deny that, since the publication of [Marat's] book of *Discoveries Concerning Light*, Newton has lost the greatest jewel in his crown."[21]

Unlike the recalcitrant Marat and others of his ilk, most savants maneuvered well within the Newtonian legacy. In 1830, the astronomer John Herschel published an influential popular account of natural philosophy, wherein he also had occasion to reflect on Newton:

A future principal player in the French Revolution and an amateur savant, Jean-Paul Marat (1743–1793) launched a campaign to excoriate the scientific establishment and to raise his standing above that of Newton, "this idol of the day." - NYPL-Print Collection

The legacy of research which Newton may be said to have left to his successors was truly immense. To pursue, through all its intricacies, the consequences of the law of gravitation; to account for all the inequalities of the planetary movements, and the infinitely more complicated, and to us more important ones, of the moon; and to give, what Newton himself certainly never entertained a conception of, a demonstration of the stability and permanence of the system, under all the accumulating influence of its internal perturbations; this labour, and this triumph, were reserved for the succeeding age, and have been shared in succession by Clairaut, D'Alembert, Euler, Lagrange and Laplace. Yet so extensive is the subject, and so difficult and intricate the purely mathematical enquiries to which it leads, that another century may yet be required to go through with the task.[22]

Herschel, as well as most earlier savants, found honor enough in proceeding along the route charted by the great Newton, even as they disagreed with, and enlarged upon, his work. Indeed, there arose a strong tradition of praising by comparison with Newton. Condorcet, for example, was certain he paid d'Alembert a great compliment in 1782 when he pronounced before the Académie Française – d'Alembert himself being present – that the greatness of that mathematical physicist consisted in no small degree of his "resolving the most difficult problems that Newton left to his successors." A decade later, in his *Esquisse d'un tableau historique des progrès de l'esprit humain* ("Sketch for a Historical Picture of the Progress of the Human Mind"), Condorcet exclaimed that "man at last discovered one of the physical laws of the universe, a law that hitherto remained unique, like the glory of the man who revealed it."[23] To an engraving of Jean-Henri Lambert was added a quatrain tying the brilliant astronomer to Newton's coattails: "Fortunate emulator of

An apotheosis of the great French naturalist Georges Louis Leclerc, comte de Buffon (1707–1788). – Burndy Library

the famous Newton, / He knew the system of this vast Universe. / By his learned writings, he illustrated his name; / In a word, he was great and formed himself." In much the same manner, a eulogist at Pierre-Simon de Laplace's funeral pronounced: "in his long and brilliant career, [Laplace] had the good fortune to complete" that which Newton "so happily attempted."[24]

Three examples from the 1720s and 1730s illustrate both the ambition that the Newtonian achievement awakened in gifted young savants to follow in the master's footsteps, as well as the homage Newton continued to command even after mathematics and mechanics were abandoned for other fields. Georges Louis Leclerc, comte de Buffon, the foremost French biologist of the eighteenth century, boasted in old age that he had discovered the binomial theorem at the age of twenty, "without knowing that it had been discovered by Newton"; he had desisted from publishing it only "because no one [was] obliged to believe [him] about it." Symbolically, his dating placed the putative discovery in 1727, the year of Newton's death. Somewhat maliciously, in his *éloge* of Buffon, Condorcet suggested that Buffon had embarked on mathematics "for no better reason than that it was the science '*à la mode*' at that time." To Condorcet's mind, Buffon's efforts "never progressed beyond those of a student," and it was "the discovery of his own ineptitude that led him at last to abandon the subject." Certainly, in retrospect there seems something calculating about the topic Buffon initially chose to establish his reputation; whatever his motivation, even after he turned from mathematics to natural history, his admiration for Newton was unwavering. He composed his masterpiece, *Histoire naturelle* ("Natural History"), in an almost entirely bare room, the sole decoration an engraving of Newton on the wall. More significantly, Buffon's conception of the origins and stability of the solar system was fundamentally Newtonian; so, too, universal gravitation served as a model for his conceptualization of the origins and nature of life.[25]

Denis Diderot, by his own admission, studied Newton carefully during the 1730s "with the intention of elucidating him." He "pushed on, if not with great success, at least with adequate vivacity," and he gave up on the venture only in 1739, upon learning that an edition of and a commentary on the *Principia* had just been published by the Minim priests Thomas Le Seur and François Jacquier. Though Diderot turned his attention elsewhere, Newton held sway over him, albeit tacitly. Notably, when rejecting rationalism as well as mathematics in his *Pensées surl'interprétation de la nature* ("Thoughts on the Interpretation of Nature") in 1753, Diderot chastised Newton for not devoting a mere month of his life to clarifying the *Principia* – a small effort that "would have spared a thousand good minds three years of labour and fatigue."

But Diderot's real target was not so much the science of the *Principia* as the rigorous mathematization, bordering on rationalism, that d'Alembert had introduced into mechanics. Indeed, Diderot singled out Newton's scientific methodology, and the physics delineated in the *Opticks*, as the yardstick for true natural philosophy. His regard for Newtonianism as the fulfillment of Baconian science was signaled by his use of the paradigmatic Newtonian optical experiment as his weapon of choice to refute rationalist philosophers. Whereas the rationalist argues magisterially "*Light cannot be split*," the genuine (Newtonian) experimental philosopher produces a prism and says, "*Light can be split*."

Noteworthy also is Diderot's publication in 1761 – in the Jesuit *Journal de Trévoux*, no less – of an article upholding the verity of the inverse square law and its applicability to various phenomena in the terrestrial domain.[26]

A somewhat surprising candidate to demonstrate Newton's pervasive hold on the minds of contemporaries is Jean-Jacques Rousseau. Like Buffon and Diderot, Rousseau also sought early on to master mathematics and Newtonian philosophy, in his case during his fruitful period of study at the manor of his mistress Mme de Warens in 1736–37. A poem Rousseau wrote shortly thereafter offers insight into the authors he read and indicates his budding philosophical preferences. In reading Leibniz and Newton, Rousseau wrote, his reason approached the sublime; Descartes, in contrast, offered only sublime aberrations and frivolous romance. Years later, Rousseau admitted that these same philosophical studies led him to embark on a "chimerical project" in which he sought to harmonize the contradictory opinions of Newton, Descartes, Leibniz, and Malebranche. He quickly gave up on the idea, but in 1738 he sought to enter the then fashionable debate over the shape of the earth. To the *Mercure de France*, he submitted a paper – which the editors declined to publish – refuting an anti-Newtonian article the periodical had published a few months earlier.

A decade later, Rousseau used his polemical *Discours sur les avantages des sciences et des arts* ("A Discourse on the Moral Effects of the Arts and Sciences") – which argued that progress in the sciences contributed little to human morality and happiness – to pronounce, perhaps with a touch of self-flagellation, that advances in physics and philosophy were begot by idle curiosity and ambition. Significantly, Newtonian science was singled out as the prime example: "Tell me then, illustrious philosophers, of whom we learn the ratios in which attraction acts *in vacuo*; and in the revolution of the planets, the relations of

This plate, from Jean-Antoine Nollet's *Essai sur l'electricité des corps*, depicts the famous Leyden jar experiment (which established that water is an excellent conductor of electricity) against the background of Nollet's theory of electrical attraction and repulsion. Jean-Jacques Rousseau, typical of a generation fascinated by the power and wonder of electricity, wrote in his *Lettres morales* (written 1757–58) of the futility of seeking knowledge of the external world, exemplified by "the great Newton, interpreter of the Universe," who yet remained ignorant of electricity. – NYPL–Rare Books Division

spaces traversed in equal times?" Elsewhere in the *Discours*, Rousseau singled out the "teachers of mankind" – Newton, Bacon, and Descartes – as proof of the futility of educating the generality of people in science; "those whom nature intended for her disciples have not needed masters."[27]

The presence of Newton in Rousseau's thought shines through in the "creed of the Savoyard priest" of *Emile*, a mirroring of Rousseau's own profession of faith. Attesting to his privileging both of the natural theology of Samuel Clarke (and Isaac Newton) and of Newtonian cosmology – albeit with his signature skepticism – Rousseau conceives of a universe made up of diffused and dead matter, wherein the heavenly bodies move according to "ordered, uniform, and ... fixed laws," with universal gravitation their unknown cause. Such a conception provides Rousseau with the building blocks for his (basically Newtonian) "first principle": "there is a will which sets the universe in motion and gives life to nature."

Rousseau's tacit acceptance of Newton as the supreme modern authority on natural philosophy is reinforced in his *Letter to Beaumont*, wherein he insists that to prove a miracle, one must first know all the laws of nature in order to ascertain the deviation from them. But since even "Newton did not boast of knowing them," Rousseau finds fodder for his relativist stance. Newton's shadow becomes even more pronounced in Rousseau's *Lettres morales* ("Moral Letters"): the futility of seeking knowledge of the external world is exemplified by "the great Newton, interpreter of the Universe," who yet remained ignorant of the wonders of electricity – the most active principle of nature as far as Rousseau was concerned. This oblique, but seminal, relationship with Newton no doubt influenced the decision of a late eighteenth-century editor of Rousseau's works to commission an engraved portrait in which the anti-Enlightenment hero looks out from his study on a pastoral landscape; on his desk sit two books, Newton's *Principia* and Locke's *Essay Concerning Human Understanding*.[28]

As the above examples suggest, the exact sciences either proved increasingly daunting for ambitious young savants, or other intellectual endeavors better suited their temperaments and inclinations. Nevertheless, many attempted to transplant the Newtonian model to other realms. Newton himself had suggested such a possibility in the preface to the *Principia*. There, he expressed the hope that the "same kind of reasoning" that had enabled him to unravel the operation of the forces of nature regulating the motion of the heavenly bodies would ultimately lead to the discovery of the forces regulating other phenomena of nature. Newton broadened his prescription for future investigation in the conclusion to the second edition of the *Opticks*, where he waxed eloquent on the contribution of his method of analysis and composition to his discovery of the nature and properties of light and colors. "By pursuing this Method," Newton predicted, natural philosophy "in all its Parts" would be perfected; then, he was confident, "the Bounds of Moral Philosophy [would] be also enlarged."[29]

Newton's recommendation encouraged philosophers, political theorists, and moralists, who came increasingly to believe that the regularity Newton had discovered in nature had its counterpart in the domain of man and society. These thinkers sought to effect in their respective fields the kind of revolution Newton had effected for natural philosophy; or, failing that, at least to avail themselves of specific Newtonian metaphors and analogies as well as the powerful new value-system that science had afforded. The Irish idealist philosopher George Berkeley was among the first to recommend such a transfer. Within a month of the publication of the second edition of the *Principia*, Berkeley published in *The Guardian* an essay, "The Bond of Society," in which he argued that the moral and intellectual domains parallel the natural and corporeal domains – for Berkeley a parallel indicative of divine purpose.

Beginning with a succinct and sound account of Newtonian celestial mechanics, Berkeley proceeded to note that "a like principle of attraction" operated "in the Spirits or Minds of men ... whereby they are drawn together in communities, clubs, families, friendships, and all the various species of society." And just as the force of attraction varies in

ALCIPHRON:
OR, THE
MINUTE PHILOSOPHER.
IN
SEVEN DIALOGUES.
Containing an APOLOGY *for the* Chriſtian Religion, *againſt thoſe who are called* Free-thinkers.

VOLUME *the* SECOND.

The Balances of Deceit are in his Hand. Hoſea. xii. 7.
Τὸ ἐξαπατᾶσθαι αὐτὸν ὑφ' αὑτοῦ πάντων χαλεπώτατον. Plato.

The SECOND EDITION.

LONDON:
Printed for J. TONSON in the *Strand*. 1732.

George Berkeley's *Alciphron: or, The Minute Philosopher* (London, 1732) was part of his powerful defense of the Christian religion against deism; the campaign also included his critique of Newton's calculus in *The Analyst*, published two years later. – NYPL–General Research Division

proportion to the mass and distance of the heavenly bodies, so in the human soul "affection towards the individuals of the same species who are distantly related to it, is rendered inconspicuous by its more powerful attraction towards those who have a nearer relation to it. But as those are removed the tendency which before lay concealed doth gradually disclose itself." This extended analogy led Berkeley to posit a tentative explanatory framework to account for human emotions (such as love) and societal relations (such as family and friendship).[30]

John Theophilus Desaguliers (1683–1744) was one of the most successful public lecturers on experimental philosophy in the first half of the eighteenth century. His public demonstrations and publications greatly facilitated the diffusion of Newtonianism both in England and on the Continent. - Burndy Library

This is not to suggest that Berkeley was a disciple of Newton. Quite the opposite. Early on he articulated powerful criticism of certain aspects of Newtonian science, and later an even more powerful critique of Newton's mathematics, in no small part owing to his concern for the implications for religion.[31] Such criticism, however, did little to diminish his esteem for Newton and, more important, for the intellectual scaffolding Newton had introduced. John Theophilus Desaguliers, on the other hand, was one of Newton's staunchest supporters, and did much to confirm Newton's experiments before the Royal Society and in other public lectures as well as through his writings. In 1728, Desaguliers commemorated the recently deceased Newton with an allegorical poem, *The Newtonian System of the World: The Best Model of Government*, in the preface of which he attributed the happiness of the British people to their life under "limited Monarchy," which made them "sensible, that ATTRACTION is now as universal in the Political, as the Philosophical World." The nearly 200 lines that followed on the progress of astronomy from antiquity to Newton – replete with lengthy technical notes – culminated in an explicit parallel between the scientific and political realms:

Boldly let thy *perfect model* be,
Newton's (the only true) *Philosophy*;
Now sing of Princess deeply vers'd in Laws,
And Truth will crown thee with a just
Applause;
Rouse up thy Spirits, and exalt thy Voice
Loud as the Shouts, that speak the People's
Joys;
When Majesty diffusive Rays imparts,
And kindles Zeal in all the *British* Hearts,
When all the Powers of the Throne we see
Exerted, to maintain our *Liberty*:
When Ministers within their Orbits move,
Honour their King, and shew each other
Love:
When all Distinctions cease, except it be
Who shall the most excell in Loyalty:
Comets from far, now gladly wou'd return,
And, pardon'd, with more faithful Ardour
burn.
ATTRACTION now in all the Realm is seen
To bless the Reign of GEORGE and
CAROLINE.[32]

The Newtonian System of the World appeared during the sojourn to England of Charles-Louis de Secondat, baron de Montesquieu. Given the friendship Montesquieu struck up with Desaguliers while in England, he was most likely familiar with the poem and its argument, and perhaps drew on it for his *De l'esprit des loix* ("The Spirit of Laws"). Educated a Cartesian, with little inclination for the mathematical and physical sciences, Montesquieu overcame his education as well as the influence of his close friend – and implacable anti-Newtonian – the Jesuit Pierre-Louis Castel, to defend Newton. Montesquieu's growing admiration for Newton can be glimpsed in his choice to carry with him on his travels a copy of Fontenelle's *éloge* of Newton, and even more so in his dismissal of a certain caviler of Newton, perhaps his friend Castel: "the person who can not create a system like Newton will make an observation with which he will torture that great philosopher; Newton, however, will always be Newton, that is to say, Descartes' successor, and the latter a common man."

De l'esprit des loix demonstrates Montesquieu's grafting of certain characteristics of scientific laws onto human and social laws. Such laws were descriptive, not normative, and were intended to reduce his observations of man's relations with nature and society to a handful of principles, which, in turn, allowed him to derive further consequences. Montesquieu also made two explicit allusions to the *Principia*. In the first, he likened the monarchical form of government to the system of the universe, "in which there is a power that constantly repels all bodies from the centre, and a power of gravitation that attracts them to it. Honor sets all the parts of the body politic in motion, and by its very action connects them; thus each individual advances the public good, while he only thinks of promoting his own interest." Later, he introduced a muted version of Newton's third law when he sought to address the relation of education to the "principles of government." "The relation of laws to this principle," Montesquieu speculated, "strengthens the several springs of government; and this principle derives thence, in its turn, a new degree of vigor. And thus it is in mechanics, that action is always followed by reaction." The naturalist Charles Bonnet

This ingenious vignette from Louis-Bertrand Castel's *Le vrai système de physique générale de M. Isaac Newton* (Paris, 1743) shows Cartesian players – using cues – victorious in a game of billiards over the Newtonians, who employ a parallelogram of forces. The further defeat of Newtonian optics is demonstrated in the background of the scene. – Courtesy of the California Institute of Technology Archives

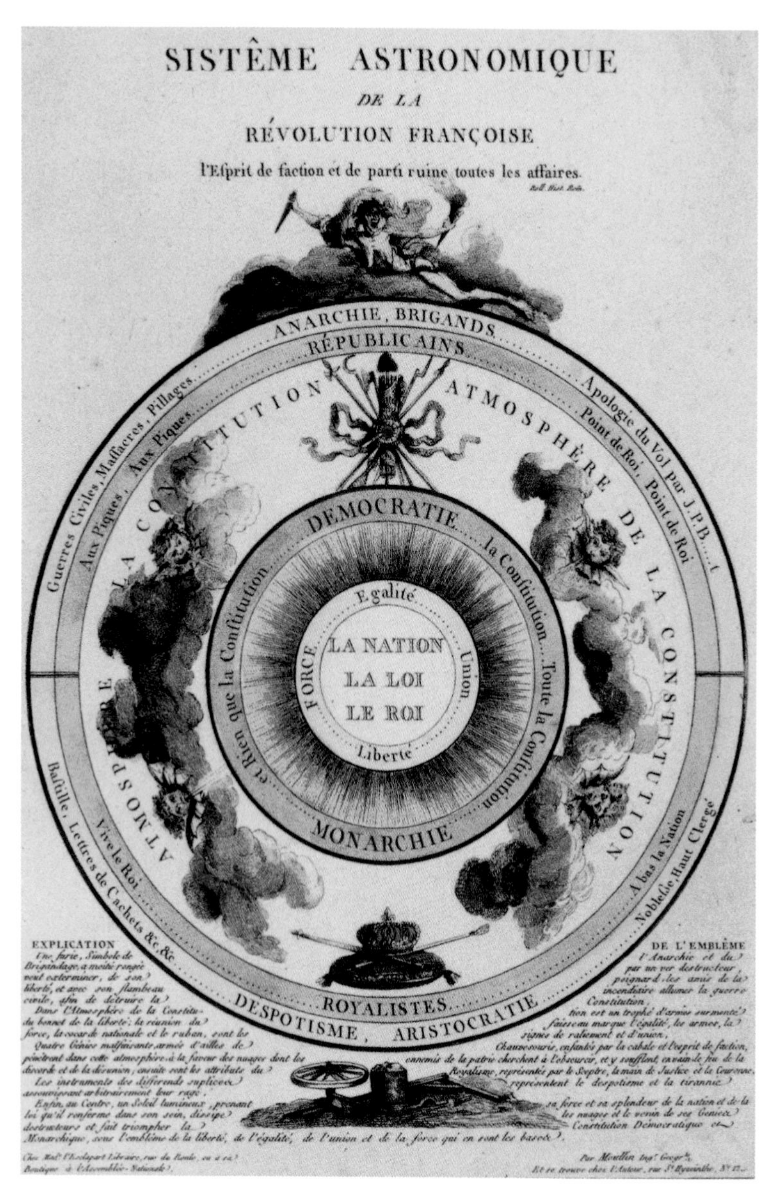

In the early phase of the French Revolution, a proponent of constitutional monarchy named Moullin was inspired by the Newtonian conception of the solar system to draw his ***Sistême astronomique de la révolution françoise***, which likened "The Nation, The Law, [and] The King" to a luminous sun dispersing the clouds, bringing about "the triumph of the Democratic and Monarchic Constitution under the emblem of the liberty, equality, union, and force which are its foundations." – Burndy Library

paid tribute to Montesquieu's *De l'esprit des loix* by drawing a parallel between his discovery of "the laws of the intellectual world" and Newton's discovery of "the laws of the material world."[33]

Montesquieu can by no means be considered a full-fledged Newtonian political theorist. Yet his tentative efforts to borrow Newtonian concepts were followed by the attempts of a flurry of younger theorists better versed in Newtonian science and more aware of the cachet it offered. Of the many political theorists who latched onto the Newtonian legacy, Thomas Jefferson is an outstanding example. His admiration of Newton was profound, matched by a more than common mastery of the Englishman's writings. Not satisfied with keeping a portrait of Newton on his wall, Jefferson obtained one of the handful of copies of Newton's death mask made in 1727 by John Michael Rysbrack. Upon retiring from the presidency in 1812, Jefferson confessed his delight in giving up "newspapers in exchange for Tacitus and Thucydides, for Newton and Euclid." Two years later he wrote a friend that "we might as well say that the Newtonian system of philosophy is a part of the common law, as that the Christian religion is."

Far more important, Jefferson's steadfast faith in the validity of Newton's laws caused him, after a fashion, to provide a Newtonian frame for the American Declaration of Independence. Thus, the invocation of "Laws of Nature" in the preamble was calculated to draw in the minds of his readers a parallel with the immutable laws of motion established by Newton. The celebrated

"We hold these truths to be self-evident" invokes the axiomatic nature of such laws vis-à-vis citizens' rights. The Newtonian (and Lockean) scaffolding introduced by Jefferson would have been obvious even to contemporaries who had not studied Newton, for by 1776 analogies between the laws of the natural world and the laws of the social world had become a fixture of public discourse.[34]

To the commentators on Adam Smith – as well as to Adam Smith himself – *The Wealth of Nations* deserved to be regarded as the *Principia* of economics. As a student at Glasgow, Smith had studied Newton carefully, and his esteem for the latter's accomplishment informed his incomplete early attempt (ca. 1758) to provide an expansive philosophical system, *The Principles Which Lead and Direct Philosophical Enquiries; Illustrated by the History of Astronomy*. Here, the certitude ascribed to the Newtonian system was contrasted sharply with the absence of certitude in earlier philosophies: "Such is the system of Sir Isaac Newton, a system whose parts are all more strictly connected together, than those of any other philosophical hypothesis." Despite opposition, Newton's system prevailed "and has advanced to the acquisition of the most universal empire that was ever established in philosophy." Newton's "principles, it must be acknowledged, have a degree of firmness and solidity that we should in vain look for in any other system." Although in his work Smith endeavored "to represent all philosophical systems as mere inventions of the imagination, to connect together the otherwise disjointed and discordant phaenomena of nature," he could not but

> make use of language expressing the connecting principles of this one [Newton's], as if they were the real chains which Nature makes use of to bind together her several operations. Can we wonder then, that it should have gained the general and complete approbation of mankind, and that it should now be considered, not as an attempt to connect in the imagination

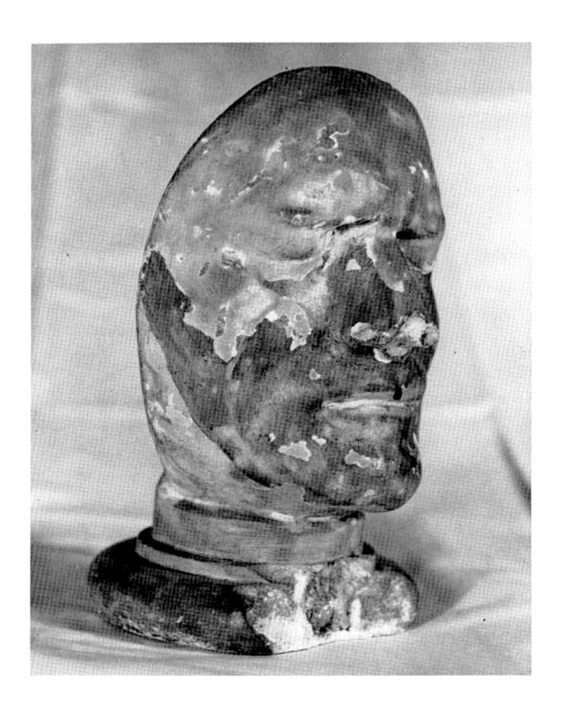

John Conduitt, who married Catherine Barton, Newton's niece, commissioned John Michael Rysbrack to prepare Newton's death mask following the great man's death on March 20, 1727. A number of plaster copies were made at the time; this one was once owned by Thomas Jefferson. – Burndy Library

> the phaenomena of the Heavens, but as the greatest discovery that ever was made by man, the discovery of an immense chain of the most important and sublime truths, all closely connected together, by one capital fact, of the reality of which we have daily experience.[35]

Smith expounded similar sentiments in his *Lectures on Rhetoric and Belles Lettres*, where he delineated the "Newtonian method": "certain principles known or proved in the beginning" are laid down "from whence we account for the severall Phenomena, connecting all together by the same Chain." Such methodology, Smith pronounced, was "the most Philosophical," both in natural and in moral philosophy. By the time *The Wealth of Nations* was published in 1776, therefore, Smith no longer felt the need to articulate the superiority of the Newtonian system, or the applicability of its laws and methodology to other domains. Nevertheless, the message was implicit, much as it was in the Declaration of Independence that same year. Smith's principles of self-interest and a self-regulating market were as foundational as universal gravitation. So, too, his economic system operated according to regulatory principles much like Newton's mechanics: "The natural price, therefore, is, as it were, the central price, to which the prices of all commodities are continually gravitating. Different accidents may sometimes keep them suspended a good deal above it, and sometimes force them down even somewhat below it. But whatever may be the obstacles which hinder them from settling in this center of repose and continuance, they are constantly tending towards it."

Smith's methodology, too, was akin to Newton's in deriving general principles from observed phenomena, and then utilizing those principles to deduce further phenomena. Smith also emulated Newton's celebrated modesty, avowing that he could not ascertain whether the principles he proposed were primary principles or manifestations of more fundamental principles still; for him, as for Newton, it was enough to recognize their universality.[36]

Other economists made similar use of Newtonian analogies and metaphors. Ferdinando Galiani, nephew of the great Italian apostle of Newton, Celestino Galiani, imbibed in his uncle's house a strong taste for Newtonian ideas, not least Celestino's confidence that "just as the forces and laws with which [the passions] operate are disclosed by Newtonian physics, and known as they are, using these as principles we cometo know other phenomena." Ferdinando made use of such a notion in his influential *Della moneta* ("On Money") of 1751, where he argued that "the laws of commerce correspond with as great exactness to those of gravity and of liquids as nothing else could. Gravity in physics is the desire in humans to earn or to live happily; that said, all the laws of physics about bodies can be perfectly verified, by whomever knows how to do so, in the morals of our life." The economist Isaac de Pinto did not apply universal gravitation to his treatise on credit, but he found occasion to invoke it when criticizing as chimeras the principles of the Comte de Mirabeau. De Pinto added: "as it is possible that, without Galileo and Kepler, Newton might never have analysed the rays of light, or thoroughly

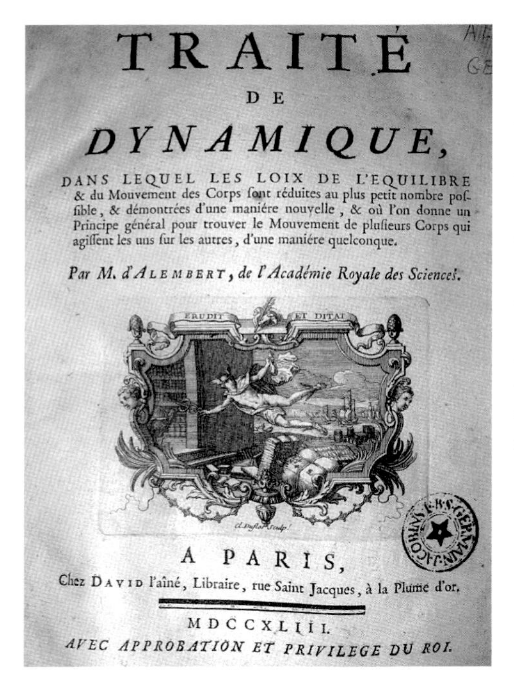
TRAITÉ
DE
DYNAMIQUE,
DANS LEQUEL LES LOIX DE L'EQUILIBRE & du Mouvement des Corps sont réduites au plus petit nombre possible, & démontrées d'une maniére nouvelle, & où l'on donne un Principe général pour trouver le Mouvement de plusieurs Corps qui agissent les uns sur les autres, d'une maniére quelconque.

Par M. d'ALEMBERT, de l'Académie Royale des Sciences.

A PARIS,
Chez DAVID l'aîné, Libraire, rue Saint Jacques, à la Plume d'or.
MDCCXLIII.
AVEC APPROBATION ET PRIVILEGE DU ROI.

In the *Traité de dynamique* (Paris, 1743–44), his most important publication, Jean Le Rond d'Alembert enunciated the principle that bears his name, which permitted the application of Newton's third law of motion to rigid bodies as well as to bodies in motion, thereby reducing a problem in dynamics to one in statics. – Courtesy of the California Institute of Technology Archives

discussed the principles of gravitation, it is to be hoped, that the work of speculative politicians will in like manner assist others in explaining the true principles, on which the finances should be administered for the welfare and happiness of mankind."[37]

Newton's influence permeated the works of the Italian philosopher and economist Antonio Genovesi. In *Della filosofia del giusto e dell'onesto* ("On the Philosophy of the Just and the Honest"), Genovesi argued that the "physical laws of the world are the basis on which moral ones rest," then substantiated the claim by drawing numerous analogies between the two domains. Like Berkeley earlier, Genovesi inferred from the fact that attraction between two bodies weakens in proportion to the distance between them that "the reciprocal attraction of men and charity is greatest in the unions of blood, shared residence, and the fatherland ... and it languishes progressively at greater distances." Elsewhere, he commented that trust is "to the civil body what the forces of cohesion and mutual attraction are to natural ones."[38]

The attempts to transfer metaphors and analogies from the natural sciences to other domains were often informed by an eagerness to discover a single principle that would account for the diversity of phenomena – much as Newton demonstrated the potency of universal gravitation to explain a vast number of observable facts. Such eagerness manifested itself even among Newton's scientific heirs, who sought to improve on the Englishman's laws by unearthing more fundamental principles still. For example, the "d'Alembert principle," announced in 1743, permitted the reduction of a problem in dynamics to one in statics; so, too, Maupertuis' 1750 "Principle of Least Action" posited a single metaphysical principle upon which all the laws of motion could be established.[39]

By the middle of the eighteenth century, the quest for such foundational principles had become increasingly evident. The philosopher Etienne Bonnot de Condillac, for example, proclaimed in his *Essai sur l'origine des connoissances humaines* ("An Essay on the Origin of Human Knowledge") that he sought "to reduce to a single principle everything concerning the human understanding." Such a principle, Condillac hastened to add, would "not be a vague proposition, nor an abstract maxim, nor a gratuitous proposition, but an inevitable experience, all the consequences of which [would] be confirmed by new

FAITH AND REASON

Newton's religiosity was proverbial, but he was also an anti-Trinitarian heretic, albeit taking care to conceal his convictions. Also concealed from public scrutiny were his extensive researches into prophecies and the history of the Church. However, Newton's presumed religious sentiments, based on the few gleanings to be had from the *Principia* and the *Opticks*, were exploited by English and Continental Protestants seeking to promote natural theology, such as William Derham [5], William Whiston [6], and Johann Jakob Scheuchzer [3]. Newton's authority was also invoked by numerous cosmogonists who followed in the footsteps of Thomas Burnet [1] and Willem Goeree [2]. Conversely, the seeming ease with which Newton's writings could be invoked by those preaching deism or materialism confirmed the suspicions of such conservative theologians as John Hutchinson that Newtonian science could lead to irreligion, if not outright atheism. Hutchinson's forceful opposition to Newtonian science was satirized in William Hogarth's "Frontis-piss" [4], which depicted the demise of most of Hutchinson's disciples (the mice) upon reading his book, their bodies washed away by a torrent of urine delivered by a lunar witch. The few surviving mice continue to attack Newton's *Principia*.

1

2

3

4

6

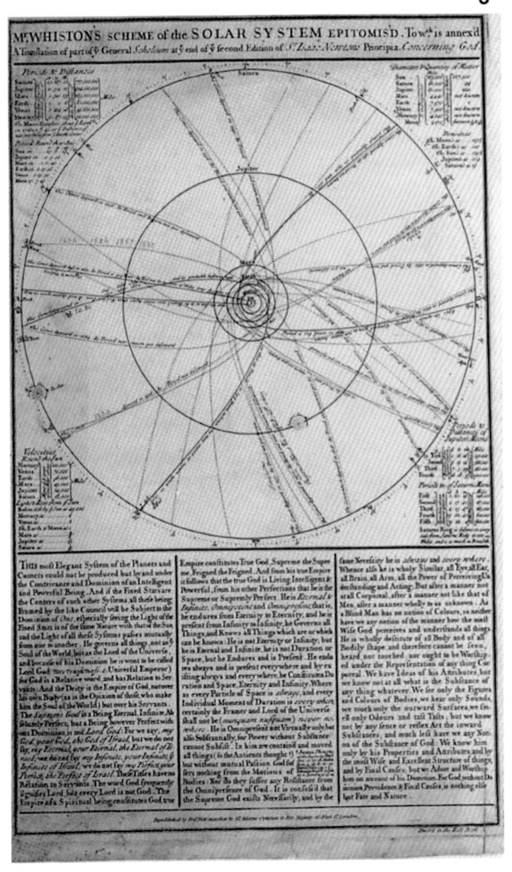

5

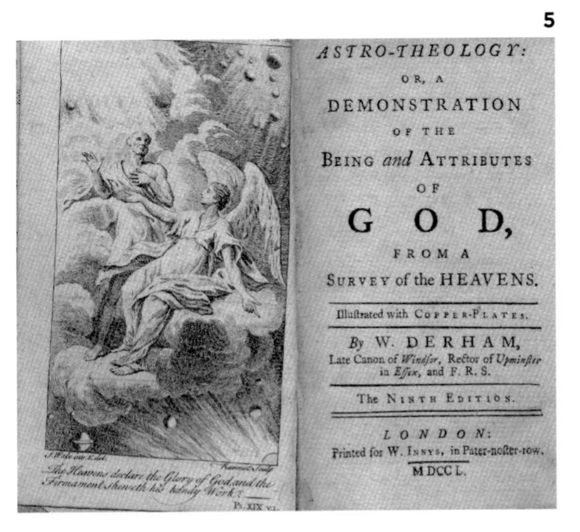
ASTRO-THEOLOGY:
OR, A
DEMONSTRATION
OF THE
BEING and ATTRIBUTES
OF
GOD,
FROM A
SURVEY of the HEAVENS.

Illustrated with COPPER-PLATES.

By W. DERHAM,
Late Canon of Windsor, Rector of Upminster in Essex, and F. R. S.

The NINTH EDITION.

LONDON:
Printed for W. INNYS, in Pater-noster-row.
MDCCL.

The Heavens declare the Glory of God and the Firmament sheweth his handy Work.

1: Thomas Burnet, *The Sacred Theory of the Earth* (London, 1684) – Courtesy of the California Institute of Technology Archives

2: Willem Goeree, *Voor-bereidselen tot de Bybelsche wysheid* (Amsterdam, 1690) – Private Collection

3: Johann Jakob Scheuchzer, *Physica sacra* (Augsburg and Ulm, 1731–35) – NYPL–General Research Division

4: William Hogarth, "Frontis-piss," 1763 – The Pierpont Morgan Library, New York

5: William Derham, *Astro-theology: or, A Demonstration of the Being and Attributes of God, from a Survey of the Heavens* (London, 1750) – Private Collection

6: *Mr. Whiston's Scheme of the Solar System Epitomis'd. To wch. is annex'd a translation of part of ye General Scholium at ye end of ye second edition of Sr. Isaac Newton's Principia. Concerning God* (London, ca. 1825) – Harvard College Library experiences." The doctrine he chanced upon was the "association of ideas," which provided him with a Newtonian principle "comprehensive enough to give a plausible explanation of our mental processes, concrete enough to be observed in action, and mechanistic enough to fit the scientific scheme of things."[40]

Even more ambitious was the Utilitarian philosopher Jeremy Bentham, whose "fundamental maxim" was encapsulated in the slogan "The greatest happiness of the greatest number is the measure of right and wrong." For his predecessors' vague accounts of utility, Bentham substituted a formula that implied the "possibility of an accurate quantitative comparison of different sums of happiness." Bentham, in fact, argued that his precise formula stood in the same relation to previous accounts as the "difference between the statement that the planets gravitate towards the sun, and the more precise statement that the law of gravitation varies inversely as the square of the distance." Small wonder that on these grounds he hoped to become the "Newton of the moral world."[41] Nor was the domain of the arts immune to such reductionism. In 1746, the French theorist Charles Batteux published a book entitled *Les Beaux-Arts réduits à un même principe* ("The Fine Arts Reduced to a Single Principle"), wherein he put forward the principle of *la belle nature*: a prototype of perfection he believed to exist behind the "reality" of things and which conditioned the mind of the artist as he sought to imitate nature. Across the channel, the young Edmund Burke hammered out a "standard of taste" whose axioms, much as in a science, were reducible "into a system." The Newtonian inspiration was obvious to such contemporaries as William Blake, who recalled that he had read the book as a young man, and instantly felt "Contempt and Abhorrence," as it was "founded on the Opinions of Newton and Locke" who "mock[ed] Inspiration and Vision."[42]

Undoubtedly, the most colorful would-be discoverer of universal principles was the French social theorist Charles Fourier. Announcing his discovery of a principle of human motivation in 1803, Fourier mused over the four apples that had changed history: "Two were famous by the disasters which they caused, that of Adam and that of Paris, and two by services rendered to mankind, Newton's and my own." His epiphany, he

In "Fourierists' Heaven," the frontispiece to volume 1 of the *Almanach phalanstérien* (1845), Jesus offers his hands to Socrates on his right and to Charles Fourier – "the man who has discovered the law of the kingdom of God that was promised to the earth" – on his left. Behind Fourier stands Newton with a sphere. – Private Collection

recalled, occurred in 1799 when he reflected in a restaurant that while an apple there cost fourteen sous, the same amount could buy a hundred apples in the country. At that moment "were born the investigations which, at the end of four years, made me discover the theory of industrial series and groups and subsequently the laws of universal movement missed by Newton." Upon making his discovery public, Fourier also announced that whereas Newton discovered the laws of gravitational attraction, it had been left to him, "'an almost illiterate shop clerk,' to advance from the realm of pure curiosity (Newton's equilibrium of stars) to that of the most urgent utility (the equilibrium of passions)." His law of "passionate attraction completed Newton's work and was destined to 'conduct the human race to opulence, sensual pleasures and global unity.'"[43]

There is perhaps no better testament to the huge shadow cast by Newton over learned culture during the Enlightenment era and beyond than the frequency with which creative individuals regarded themselves (or were regarded by others) as the Newtons of their respective realms. (Rarely, if ever, did anyone wish to become a second Descartes or Leibniz.) To the many examples already cited should be added David Hume, the Newton of the moral sciences – a designation also appropriated by Kant for Rousseau; David Hartley, the Newton of the mind; Christian Wolff and Immanuel Kant, the Newtons of metaphysics; and Jean-Philippe Rameau – and, later, Johann Sebastian Bach – the Newtons of music. For the Swiss man of letters Isaac Iselin, François Quesnay's *Ephémérides du Citoyen* made the economist "appear to him what Newton [was] to a mathematician." Most endearing of all, the legal theorist Cesare Beccaria was thrilled to hear friends refer to him as "little Newton" ("il newtoncino").[44]

7

Giovanni Battista Pittoni was one of the earliest artists to apotheosize Newton. However, his "Allegorical Monument to Sir Isaac Newton" (1727–29) depicted the famous prismatic experiment inaccurately: Pittoni bent the ray of light entering the prism as well as reversed the order of the orange and yellow hues of the spectrum. – © The Fitzwilliam Museum, University of Cambridge

APOTHEOSIS

Newton died on March 20, 1727. Eight days later, he lay in state in the Jerusalem Chamber of Westminster Abbey, where he was given a public funeral the likes of which no layman had been previously accorded. Two dukes, three earls, and the Lord Chancellor supported the pall; the bishop of Rochester officiated over the services. Newton also became the first man of science to be buried at the Abbey. The funeral deeply impressed Voltaire, who experienced the spectacle firsthand. Several years later, he commented that Newton "lived honoured by his compatriots and was buried like a king who had done well by his subjects."[1] The tomb became something of a London tourist attraction for generations of admirers.

No sooner was Newton laid to rest than John Conduitt – friend, biographer, and husband of Newton's niece – began to plan an appropriate memorial. Four years and several designs later, the marble monument was unveiled at the Abbey. Designed by the architect William Kent and executed by the sculptor John Michael Rysbrack, the monument now stands against the choir screen, to the north

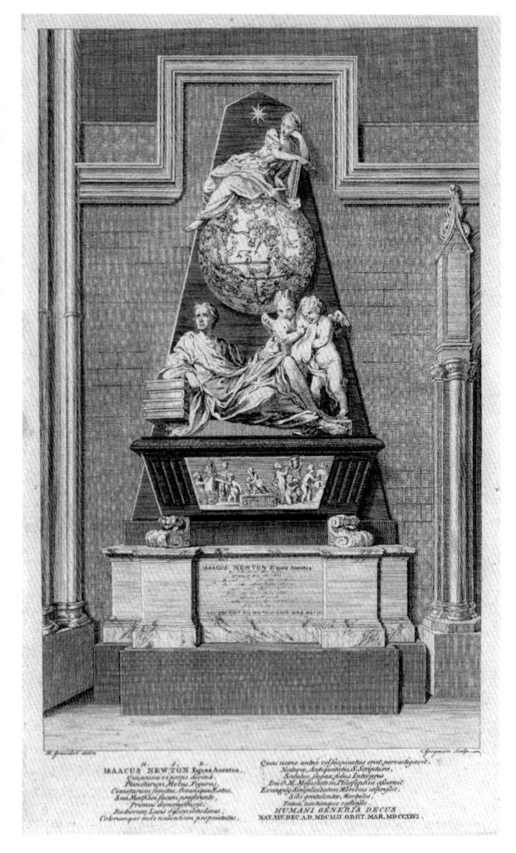

Newton's marble tomb and monument in Westminster Abbey was designed by architect William Kent and sculpted by John Michael Rysbrack. - Private Collection

of the entrance to the choir. Much in the neoclassical tradition, a robed, rather muscular Newton reclines on a sarcophagus, his right elbow propped up on a stack of four books, corresponding to his (mostly unpublished) contributions to chronology and theology as well as the *Principia* and the *Opticks*. Two angels hover, holding between them a scroll on which is displayed a diagram of the solar system. A large celestial globe dangles overhead, its surface configured with the putative zodiacal position of the constellations at the time of the Argonauts' expedition (established by Newton in his chronology) as well as with the 1680 comet whose orbit Newton also computed. A forlorn Lady Astronomy presides in mourning atop the globe; on the bas relief decorating the sarcophagus, several putti wield scientific instruments associated with Newton's scientific work: a telescope, a prism, a steelyard, and a furnace. The engraved inscription, in Latin, reads:

Here is buried Isaac Newton, Knight, who by a strength of mind almost divine, and mathematical principles peculiarly his own, explored the course and figures of the planets, the paths of comets, the tides of the sea, the dissimilarities in rays of light, and, what no other scholar has previously imagined, the properties of the colours thus produced. Diligent, sagacious and faithful, in his expositions of nature, antiquity and the holy Scriptures, he vindicated by his philosophy the majesty of God mighty and good, and expressed the simplicity of the Gospel in his manners. Mortals rejoice that there has existed such and so great an ornament of the human race![2]

By the time the monument was unveiled, the apotheosis of Newton was already in full sway. Indeed, the process of deification may be said to have begun with the meeting of the Royal Society on April 28, 1686, when a copy of Book I of the *Principia* was presented to the Fellows. After its contents had been described and its novelty extolled, one Fellow exclaimed that with his book Newton had carried the subject so far, "that there was no more to be added." To which the Vice-President retorted from the chair, "that it was so much the more to be prized, for that it was both invented and perfected at the same time." Edmond Halley, who did so much to secure the publication of the *Principia*, bestowed on Newton even higher praise in his ode prefacing the volume, the final few verses of which read:

O you who rejoice in feeling on the nectar
of the gods in heaven,
Join me in singing the praises of Newton,
who reveals all this,
Who opens the treasure chest of hidden
truth,
Newton, dear to the Muses,
The one in whose pure heart Phoebus
Apollo dwells and whose mind he has
filled with all his divine power;
No closer to the gods can any mortal rise.[3]

Four decades later, in the aftermath of Newton's death, John Conduitt expressed much the same sentiments, if less poetically: "had this great and good man lived in an age when those superior Genii inventors were Deified or in a country where mortals are canonised he would have had a better claim to these honours than those they have hitherto been ascribed to."[4] Newton also happened to be an Englishman, and his death set off a literary outpouring, with poets vying to outdo each other with expressions of admiration and grief. Perhaps the best, and most successful, of the eulogists was James Thomson, whose *To the Memory of Sir Isaac Newton* was published within three months of Newton's death and became the model for others to emulate. The poem opened with a rhetorical question studded with classical tropes:

Shall the great soul of Newton quit this
earth,
To mingle with his stars; and every Muse,
Astonish'd into silence, shun the weight
Of honours due to his illustrious name?

James Thomson's Newtonianism was evident in his magnum opus, *The Seasons* (1726–30), which exerted considerable influence on eighteenth-century English and Continental "scientific" poetry. This plate (from the 1730 London edition) is the frontispiece to "Spring." – NYPL–Rare Books Division

As if there was any doubt of the immortality that was his due, Newton, the "All-piercing sage," the "beloved of heaven," was depicted as the instrument of "the secret hand of Providence," to whom was revealed universal gravitation, the cause of the tides, the elliptical orbit of comets, the spectrum and the rainbow, and who revealed them all, in turn, to admiring humanity.[5]

A lesser poet, Richard Glover, extolled "Newton's genius, and immortal fame" in a lengthy poem he contributed to Henry Pemberton's *A View of Sir Isaac Newton's Philosophy* (1728). In line with what had by now become a tradition, Glover tied Newton to the realm of the divine – the first to display the "almighty's works" – then enumerated these works, from universal gravitation – "by whose simple power / The universe exists" – to tides, comets, and light and colors. Glover concluded:

And lo, th' all-potent goddess *Nature* takes
With her own hand thy great, thy just
reward
Of immortality; aloft in air
See she displays, and with eternal grasp
Uprears the trophies of great Newton's
fame.[6]

The Scottish poet Allan Ramsay, too, sought consolation for his grief in Newton's imminent immortality:

The god-like man now mounts the sky,
Exploring all yon radiant spheres;
And in one view can more descry
Than here below in eighty years.

And again, the poet and dramatist Aaron Hill consoled himself with the permanence of Newton's laws and fame until that day –

The experiment depicted in the center of this vignette – from Henry Pemberton's *A View of Sir Isaac Newton's Philosophy* (London, 1728) – is that of measuring the resistance of a pendulum in water, while the instrument to the left is an apparatus to measure the collision of bodies. – NYPL–SIBL

predicted by Newton himself – when the world itself would draw to an end:

when the Suns he lighted up, shall
fade,
And all the worlds he found, are still
decay'd;
Then void, and waste, Eternity shall lie,
And Time, and Newton's name, together
die.[7]

Idolization of Newton was not confined to his compatriots. Already at the turn of the eighteenth century, the French mathematician the Marquis de l'Hôpital reputedly exclaimed: "I cannot believe otherwise than that [Newton] is a *genius*, or a *celestial intelligence* entirely disengaged from matter," which led him to wonder whether the Englishman ate, drank, or slept "like other men." Some eight decades later, in 1771, Louis-Sébastien Mercier found occasion in his utopian *L'an 2440* (translated as "Memoirs of the Year 2500") to ponder the sublimity of Newton's soul and its well-deserved immortality: "The soul of Newton has flown, by its native vigour, over all the worlds that it once weighed. It would be unjust to suppose that death had power to extinguish that mighty genius. Such a destruction would be more afflicting, more inconceivable, than that of the whole material universe."[8]

Those loath to spin images evocative of the divine settled on the term "genius." "In Newton this island may boast of having produced the greatest and rarest genius that ever arose for the ornament and instruction of the species," wrote David Hume in his *History of England* (1759–62). Voltaire concurred. The progress in the arts and sciences, he wrote in *Le siècle de Louis XIV* ("The Age of Louis XIV," 1751), was owed to but "a few geniuses scattered in small numbers in various parts of Europe," chief among them Newton, "the first to see the light," who would be studied and remembered long after the memory of rulers and statesmen had faded.[9]

Traditionally, the concept of genius had been reserved for artists and poets. Largely owing to the towering example of Newton, however, in the course of the eighteenth century the concept was redefined. Edward Young, in his *Conjectures on Original Composition* (1759), for example, developed an influential conception of genius, which attached a premium to "great originals." And though he avoided discussing philosophy and the natural sciences, when he came to enumerate instances of great English "originals," he had little doubt that Newton belonged in the company of Bacon, Shakespeare, and Milton.[10]

At greater length, William Duff, in *An Essay on Original Genius* (1767), attributed genius to imagination and originality, both indicative of "a general talent, which may be exerted in any profession." The difference was that in philosophy, "the empire of Imagination, and consequently of Genius, [was] in some degree necessarily restricted," whereas in poetry, imagination was "altogether absolute and unconfined." Duff proceeded to single out Descartes as an "original" who "struck out a path for himself ... and though he could not pursue it through its several windings, he pointed out the track which has been followed by others." Nature had conferred on him a strong imagination, and precisely for this

PORTRAITS OF ISAAC NEWTON

Only monarchs, and perhaps a few noblemen, surpassed Newton in the number of times they commissioned portraits of themselves. Indeed, fourteen portraits of Newton were executed in the twenty-five years that preceded his death in 1727. Here is a selection, all painted in that final quarter century of the great man's life.

1

1710

2

1712

3

1717

4

1718

5

1725

6

7

1727

1: Sir James Thornhill – Trinity College, Cambridge

2: Sir James Thornhill – Courtesy of the Trustees of the Portsmouth Estate

3: Charles Jervas – © The Royal Society

4: Thomas Murray – Trinity College, Cambridge

5: John Vanderbank – Trinity College, Cambridge

6: after Enoch Seeman – National Portrait Gallery, London

7: John Vanderbank – © The Royal Society

In this teeming frontispiece to the second edition of Ephraim Chambers's *Cyclopaedia; or, An Universal Dictionary of Arts and Sciences* (London, 1738), a bust of Newton serenely graces the right-hand corner of the school's upper balustrade. – The William Andrews Clark Memorial Library, University of California, Los Angeles

reason his errors represented examples of "imagination too freely indulged, and too little subjected to the prudent restraints of Judgment." The "great" Newton, in contrast, "was doubtless in Philosophy an original Genius of the first rank," whose manifold "and stupendous discoveries" in natural philosophy were "the most astonishing efforts of the human mind; and while they show[ed] the prodigious compass of that imagination, which could frame and comprehend such sublime conceptions, they at the same time clearly evince[d] the profound depth and penetration and strength of reason, which, by a kind of divine intuition, could discern and demonstrate their truth."[11]

Alexander Gerard, too, in his *Essay on Genius* (1774), insisted on the prerequisite of originality and a proper mix of imagination and judgment. "Genius is properly the faculty of *invention*; by means of which a man is qualified for making new discoveries in science, or for producing original works of art." A man of genius was one endowed with a wide-ranging and perceptive imagination that allowed him to connect ideas creatively, irrespective of their apparent difference from each other. With this as his yardstick, Gerard advanced a two-pronged conception of genius for the arts and for the sciences, with judgment and imagination as prerequisites of both. A great admirer of Bacon, Gerard believed that the celebrated statesman of science had already sketched the contours of the philosophy that Newton came to realize "perfect[ly] and accurately finished." Though he struggled to determine the greater genius, Gerard ultimately opted for Newton, whose natural philosophy, even leaving aside his original mathematical investigations, seemed "to demand an acuteness and compass of invention, which we might pronounce adequate to all the investigations of Bacon." Endowed equally with judgment and imagination, Newton was primed to make his great discoveries in mechanics as well as optics: "It was perhaps his perceiving by sense a stone or an apple fall to the ground, without any visible force impelling it, or the remembrance of this common appearance, that excited his genius, and directed it to that train of thought which conducted him at last to the investigation of these laws."[12]

Across the channel, Diderot broadened the conception of genius to include the experimental physicist – which, as noted in chapter 6, was in no small part modeled on Newton – his rationale being that the faculties of one endowed with such *génie de la physique expérimentale* were a gift of nature, not acquired, and thus not fundamentally different from the inspiration of an artist. For his part, in *De l'esprit* ("On the Mind"), Claude Adrien Helvétius argued that genius was not diminished by the contribution of predecessors to its inception. "Kepler discovered the laws by which bodies ought to gravitate towards each other," he wrote; "Newton, by the happy application of this to the heavenly bodies, which a very ingenious calculation enabled him to make, confirmed the existence of these laws. Newton, therefore, lived in a proper period, and was placed in the rank of men of genius." Elsewhere in the treatise Helvétius invoked Newton's singularity by raising the rhetorical question: "Who, besides Newton, in the last age, fixed the laws of gravitation?"[13]

Immanuel Kant, in contrast, who accepted Newton's genius as a given throughout his early works, came to reverse himself in his *Kritik der Urteilskraft* ("Critique of Judgement," 1790). Writing against the backdrop of incipient German romanticism, Kant sought to exclude from the domain of genius any activity involving learning. For him, even one who wove "his own thoughts" or who brought new ideas to a science ultimately accomplished "something that *could* have been learned." Consequently,

> all that *Newton* has set forth in his immortal work on the Principles of Natural Philosophy may well be learned, however great a mind it took to find it all out, but we cannot learn to write in a true poetic vein, no matter how complete all the precepts of the poetic art may be, or however excellent its models. The reason is that all the steps that Newton had to take from the first elements of geometry to his greatest and most profound discoveries were such as he could make intuitively evident and plain to follow, not only for himself but for every one else.[14]

Such a theoretical formulation aside, the very selection of Newton to illustrate the rationale for excluding philosophers and men of science from the ranks of genius indicates how closely Newton's name had become allied with the concept of genius in the course of the century.

The references to Newton's genius are difficult even to tally. Be that as it may, even a cursory survey of the literature demonstrates the unanimity with which contemporaries routinely regarded Newton as the embodiment of genius. When an Englishman visited Voltaire at Ferney in 1776, the octogenarian man of letters pointed at a bust of Newton in his parlor and exclaimed: "it is the greatest genius that ever existed: if all the geniuses of the universe were assembled, he should lead the band."[15] Voltaire was partisan, no doubt, but his adulation differed only in degree. The mathematician and physicist Leonhard Euler, though critical of several key Newtonian tenets – both in optics and mechanics – nevertheless included the Englishman among the greatest geniuses the world had ever seen. The French Romantic writer François-René duc de Chateaubriand viewed Newton both as a "genius" and a "divine" man. The twenty-one-year-old German dramatist and poet Friedrich von Schiller singled out Newton as the acme of human perfection: "Man had to be an animal before he knew that he was a spirit; he had to crawl in the dust before he ventured on the Newtonian flight through the universe." Samuel Johnson attributed Newton's

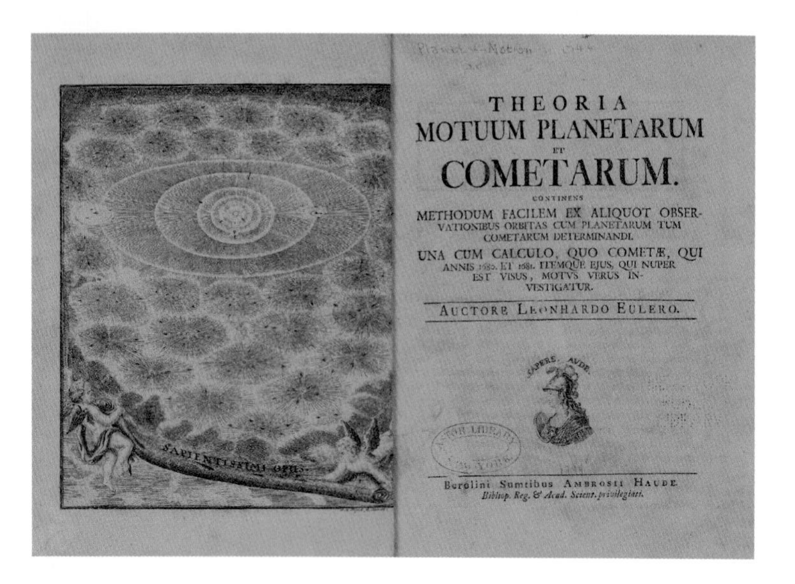
THEORIA
MOTUUM PLANETARUM
ET
COMETARUM.
CONTINENS
METHODUM FACILEM EX ALIQUOT OBSERVATIONIBUS ORBITAS CUM PLANETARUM TUM COMETARUM DETERMINANDI.
UNA CUM CALCULO, QUO COMETÆ, QUI ANNIS 1680. ET 1681. ITEMQUE EJUS, QUI NUPER EST VISUS, MOTVS VERUS INVESTIGATUR.
AUCTORE LEONHARDO EULERO.
Berolini Sumtibus AMBROSII HAUDE
Bibliop. Reg. & Acad. Scient. privilegiati.

The frontispiece to his *Theoria motuum planetarum et cometarum* (Berlin, 1744) displayed Leonhard Euler's continued adherence to a modified Cartesian theory of vortices. – NYPL–SIBL

"superiority to the rest of mankind" to more than his single-minded focus on science. Newton "stood alone, merely because he had left the rest of mankind behind him, not because he deviated from the beaten track."[16]

Perhaps the most interesting, if extreme, conception of Newton's genius was spun by the physiognomist Johann Caspar Lavater in his inventive treatise on phrenology. Even the six engravings of the Englishman available to him, Lavater announced, had made it evident that Newton was "a great" and "extraordinary man." A rendition of one painting (by Enoch Seeman) appeared to him to exhibit "the inner strength to grasp a matter, to seize upon it and not just to illuminate it, not to pile it up in the memory, but to devour and digest it and allow it to emanate in the great universe of his head." So, too, Newton's eyes were "full of the power of creation," his eyebrows "infused with the most effulgent, solid fertility." Another engraving – an idealized rendering of Godfrey Kneller's 1689 painting (the painting is reproduced on page 28, above) – made Lavater exclaim: "A man of knowledge and deed." The forehead is of one "clinched in contemplation of the workings of the universe! The severity of future tribulations is foretold in his present strength." The eyebrows are those of "the creator of new systems!" Lavater recognized that all the engravings he studied were "copies of copies"; he nonetheless believed he could look into all of them and see "men who [would] leave their mark upon future generations."[17]

While Lavater's utter confidence in the ability of engravings to convey the inner workings of Newton's soul was unusual, acquisitions of Newton's likeness were not. From the early years of the eighteenth century, when Newton's likeness first became available in the form of engravings, growing numbers of savants purchased them to adorn their studies, eager to pay homage – and perhaps to derive from them some inspiration. Indeed, Newton himself offered such engravings tovisitors. When the Italian astronomer Francesco Bianchini visited England in 1713, for example, he received one; so, too, did the French astronomer Joseph-Nicolas Delisle when he visited London a decade later. Newton's French ally Pierre Varignon actually received from him in 1720 a gift of an oil painting, executed by Kneller. In his effusive letter of thanks, Varignon divulged that he had already obtained an engraving of Newton from an English friend a decade earlier, but since an engraving could hardly represent a true likeness, he had been eager to acquire a painting. With one in hand, he could at last "see that most famous and learned man whom I have respected for more than 30 years with the greatest veneration." Nor was the desire to obtain Newton's likeness reserved for his partisans. In 1721, Newton's often bitter opponent Johann Bernoulli also sought – through Varignon's offices – to obtain one. Newton responded that he would gladly send Bernoulli a copy, if the Basel professor would acknowledge Newton's priority in discovering the calculus.[18]

Unfortunately, information regarding ownership of engravings is extremely sketchy, owing to their relatively ephemeral nature. Still, as noted earlier, we know that a portrait of Newton hung in Buffon's study at Montbard as well as in the studies of Thomas Jefferson, Immanuel Kant, and Jean Le Rond d'Alembert. John Watt, father of the inventor of the steam engine, kept in his humble house engravings of Newton and John Napier (the inventor of logarithms) that may very well have inspired the young James Watt.[19] An engraved likeness of Newton – along with one of Copernicus – graced the study of the architect Etienne-Louis Boullée, who will be discussed more fully below.

One of the more telling prints was George Bickham's "A Glorification of Sir Isaac Newton" (1732). Appropriating a motif traditionally associated with royalty – Louis XIV in particular – Bickham placed a likeness of Newton at the center of a multi-rayed blazing sun dispelling the engulfing clouds to display the solar system. An assortment of angels, muses, and accompanying putti settle mostly on clouds above and below, their arms busy with scientific instruments, their gaze admiring of Newton and his system. One owner of Bickham's engraving was the Bostonian Edward Bromfield, an ingenious optician and microscopist who died in 1746, at age twenty-three. Further evidence of his veneration of Newton is the youthful portrait of himself he commissioned; he is shown pointing at a microscope he had designed, a copy of Newton's *Opticks* on the bookshelf by his side.[20]

Busts of various shapes and substance also began to appear in increasing numbers, supplementing (or supplanting) engravings. Such busts often adorned the numerous "temples" that mushroomed in aristocratic gardens during the eighteenth century. Queen Caroline, for example, had a grotto erected for herself in the hermitage at Richmond Castle, where she meditated on natural philosophy and natural theology in the company of Newton, John Locke, Robert Boyle, William Wollaston, and Samuel Clarke.[21] At Stow, Lord Cobham built a temple of British worthies, sixteen in number, with Newton, Locke, and Bacon representing the life of the mind. At Ermenoville, in the outskirts of Paris, René-Louis Marquis de Girardin constructed in 1777 a "Temple of Philosophy," each of whose six columns was "dedicated to the memory of a great man who was useful to his fellow men by their writings and discoveries." In Girardin's printed

George Bickham's "The Apotheosis of Sir Isaac Newton" was first published in 1732. In 1787, his son republished the engraving to commemorate the centenary of the publication of the *Principia*. – Burndy Library

In "A Performance of 'The Indian Emperor; or, The Conquest of Mexico by the Spaniards,'" William Hogarth succeeded brilliantly in capturing the dramas of gravity and motion being enacted under the watchful gaze of Newton, present in the form of a white marble bust on the fireplace mantel. Robert Dodd's engraving of Hogarth's painting was published in London in 1791. – The Pierpont Morgan Library, New York

description of the temple, Newton came first with his contribution of "light," followed by Descartes ("no void in nature"), Voltaire ("ridicule"), William Penn ("humanity"), Montesquieu ("justice"), and Rousseau ("nature").[22]

While the display of engravings and busts was intended to convey an intellectual kinship with Newton, William Hogarth's "A Performance of 'The Indian Emperor; or, The Conquest of Mexico by the Spaniards'" was intended to indicate a more intimate kinship as well. It was commissioned in 1732 by John Conduitt and his wife, Catherine Barton – Newton's niece – whose painted portraits are reproduced on the wall to the left of the drawing room fireplace. The canvas, in theme and symbol, signals their intimate relationship with the great man, present in the form of a white marble bust on the fireplace mantel, under whose watchful gaze the dramas of gravity and motion unfold. On a stage to the right, the children of the Conduitts and their guests perform a scene from John Dryden's play *The Indian Emperor*, the subject of which is Hernán Cortés's conquest of Mexico and the fatal attraction to him of two Aztec princesses. In the foreground, a mother instructs her daughter to pick up a fan the girl has dropped; the implicit narrative, reinforced by the gestures of mother and daughter, invokes not only the effects of gravity but Newton's third law of motion: "For every action, there is an equal and opposite reaction." Also present, with his back to the viewer, is Newton's former Curator of Experiments at the Royal Society, John Theophilus Desaguliers, whose apparent role of prompter to the drama

In 1766, while serving in London as the representative of the American colonies, Benjamin Franklin (1706–1790) had his portrait painted by David Martin, opting to have himself depicted as if inspired by the great Isaac Newton, whose towering bust sits on the desk beside him. – White House Historical Association (White House Collection)

being enacted by the children is suggestive of "action at a distance."[23]

Three colonial American savants illustrate the various ways in which Enlightenment men of science and letters attempted to visually convey a kinship with Newton. Benjamin Franklin's life-long admiration of Newton is well documented. Upon first arriving in England in 1725, at age nineteen, he attempted to meet the great man, but without success. In subsequent years he often expressed his admiration for Newton, and regarded his work as part of the Newtonian tradition. A measure of Franklin's esteem can be gleaned from his devoting his *Poor Richard's Almanac* for December 1749 to a commemoration of Newton's birthday, although his homage to the "prince" of modern science was tinged with Alexander Pope's intellectual pessimism: "But what is all our little boasted knowledge, compar'd with that of angels? If they see our actions, and are acquainted with our affairs, our whole body of science must appear to them as little better than ignorance; and the common herd of our learned men, scarce worth their notice." "Now and then," he continued, "a Newton, may, perhaps, by his most refined speculations, afford them a little entertainment, as it seems a mimicking of their own sublime amusements." In 1766, while serving as representative of the American colonies in London, Franklin had his portrait painted by David Martin. The image he chose to convey was not just that of a man of science, deep in contemplation, but one inspired by the great Isaac Newton, whose towering bust rests on the desk beside him.[24]

Whereas Franklin chose to commemorate his kinship with Newton rather than his own scientific work – except by implication – his

A man of science as well as lieutenant governor of New York, Cadwallader Colden (1688–1776) believed that his *An Explication of the First Causes of Action in Matter* (1745) succeeded in accounting for the cause of gravity, which had eluded even Newton. The book is seen open on the desk by Colden's arm in Matthew Pratt's "Portrait of Cadwallader Colden and His Grand-daughter" (ca. 1772); a copy of the *Principia* is on the bookshelf behind him. – Private Collection

friend Cadwallader Colden sought to legitimize his own work by invoking Newton. A man of science as well as lieutenant governor of New York for fifteen years, Colden was over eighty years old by the time (ca. 1772) Matthew Pratt painted his portrait in the company of his granddaughter. By then, Colden had been working for decades on what he believed to be his crowning achievement: the discovery of the cause of gravity, which had eluded even the great Newton. He published his preliminary thoughts on the subject in 1745 under the title *An Explication of the First Causes of Action in Matter; and of the Cause of Gravitation*, with an expanded version appearing five years later. Contemporaries not only dismissed Colden's theory as obscure (and metaphysical), but generally perceived it as an attempt to overthrow Newton – a charge Colden repeatedly denied. The scathing criticism he received did little to shake his confidence in the verity of his theory, and he remained convinced until his death that if he could only further clarify it, doubters would come around. This hubris is reflected in the portrait's composition. Colden's left arm

Avid man of science, admirer of Newton, and President of Yale College, Ezra Stiles (1727–1795) instructed artist Samuel King to embody several Newtonian themes – such as an emblem of the solar system and the orbit of comets – prominently in his portrait (1771). – Yale University Art Gallery

rests on his desk, on which sits a copy of the *Principles of Action in Matter*, prominently open next to an inkwell. On the bookshelf behind him, one of the handful of books is a copy of the *Principia*, suggesting both inspiration and transcendence. (A companion painting of Colden with his grandson, probably from the same year, shows him posed with his left hand on a diagram, presumably from *Principles of Action in Matter*.)[25]

Ezra Stiles was a polymath as well as a clergyman and president of Yale College. Unlike Franklin and Colden, however, he was not the author of scientific works, and his decision in 1771 to instruct the painter Samuel King to embody several Newtonian themes in his portrait attests not only to Stiles's scholarly interests, but to the congruence he envisaged between Newtonian science and Christian religion. Stiles himself provided an explication of the portrait's meaning:

> The Effigies sitting in a Green elbow Chair, in a Teaching Attitude, with the right hand on the Breast, and the Left holding a preaching Bible. Behind & on his left side is a part of a Library – two Shelves of Books.... At my right hand stands a Pillar. On the Shaft is one Circle and one Trajectory around a solar point, as an emblem of the Newtonian or Pythagorean System of the Sun & Planets & Comets. It is pythag. so far as respects the Sun & revolving Planets: it is Newtonian so far as it respects the Comets moving in parabolic Trajectories, or long Ellipses whose Vertexes are nigh a parab. Curve. At the Top of the visible part of the Pillar & on the side of the Wall, is an Emblem of the Universe or intellectual World.

Apropos a copy of Newton's *Principia* displayed among other books on the shelf, Stiles revealed: "I posess & have read all Newtons Works & his Principia often; and am highly delighted with his Opticks & Astronomy."[26]

In contrast to these manifestations of personal interest and kinship, the apotheosis of Newton also took the form of allegory. "An Allegorical Monument to Sir Isaac Newton," for example, began as a commercial project conceived by the Irish entrepreneur Owen McSwiny to commission from Venetian painters a series of allegories to commemorate British worthies. Giovanni Battista Pittoni, assisted by the brothers Domenico and Giuseppe Valeriani, painted the allegory in honor of Newton in the late 1720s. Classical in theme, with marked rococo features, a massive, pillared Temple (or perhaps an Academy) is the site of numerous vignettes. Minerva escorts the mourning muses of the sciences up a set of stairs toward a giant urn, which seems to contain Newton's ashes. On the pedestal opposite, Mathematics, in semi-recline against a tablet, points a compass at the urn, while Truth, a torch in hand, stands over her. Crowding the floor around them, philosophers engage in a bustling study of mathematics and astronomy. The focus of all the vignettes, however, is Newton's celebrated experiment demonstrating the heterogeneity of sunlight by splitting a ray into its constituent colors. The ray of light enters from a hole on high and is captured by a mirror that reflects it into a prism. However striking this portrayal, the depiction of the experiment is inaccurate: Pittoni bent the ray of light entering the prism and reversed the order of the orange and yellow hues of the spectrum.

Painted with the assistance of Domenico and Giuseppe Valeriani, Giovanni Battista Pittoni's "An Allegorical Monument to Sir Isaac Newton" (1727–29) is rich in incident, but its focus is Newton's celebrated experiment demonstrating the heterogeneity of sunlight. – © The Fitzwilliam Museum, University of Cambridge

Whereas Pittoni's allegory attempted to incorporate an actual Newtonian experiment, Januarius Zick's twin allegories on the contribution of Newton to universal gravitation and to optics are purely symbolic in their celebration of genius. In "Allegorie auf Newtons Verdienste um die Gravitationslehre" ("Allegory on Newton's Contribution to the Theory of Gravity"), Zick literally sets Newton down on Parnassus, among the resident nine muses. The chain bound to Newton's left wrist is held by Urania, the muse of Astronomy; his left foot is chained to an iron sphere, symbolizing Earth; his right hand, gripping a compass he'd used to measure the globe, has fallen to his side, as if to indicate that his work is done: the unification of the terrestrial and celestial worlds under a single central force – symbolized by the chains and the sphere – has been achieved, and Newton is ready to receive the cup of nectar extended to him. The harmony Newton had introduced into nature corresponds to the harmony produced by the music made by the muses, implying universal harmony. On a promontory overlooking the scene below, the winged Pegasus – symbol of poetic genius – appears to rear up, suggesting an affinity between Newton's genius and that represented by the resident muses of Parnassus.

"Allegorie auf Newtons Verdienste um die Optik" ("Allegory on Newton's Contribution to Optics") even more stridently asserts Newton's genius. Represented here as a younger man stamping his foot contemptuously on the personification of falsehood (identified by the mask in his hand), Newton rests his right hand on his chest, gesturing with his left toward Euclid and Diogenes, as if to declare he has transformed the science and philosophy they represent. Towering above the two ancient philosophers are a large obelisk and a sundial, which announces the first hour of the new era. From a massive backdrop of black clouds, lightning strikes, calling to mind Newton's discoveries on the nature of light – and perhaps his gift to mankind, analogous to Prometheus's gift of fire. Through a second opening in the clouds, the sun illuminates a small temple by Newton's side. Within stands Cupid, holding out temptation; beside him is a table covered with gold and jewels. The mirror above the jewels contains Newton's reflection, perhaps alluding to his rejection of love and riches in his dedication to the search for truth.[27]

In Januarius Zick's "Allegory on Newton's Contribution to the Theory of Gravity" (ca. 1790), Newton is ready to receive the cup of nectar offered him in Parnassus, as his work is done: he has unified the terrestrial and celestial worlds under a single central force, symbolized by the chains and the sphere.

In a companion piece, below, the "Allegory on Newton's Contribution to Optics" (ca. 1790), Zick glorifies Newton's discoveries on the nature of light and colors. Symbolically, the large sundial above the two ancient philosophers proclaims the first hour of the new era. – Niedersächsisches Landesmuseum Hannover

The service rendered by Newton to mankind is the theme, on a much larger scale, of James Barry's "Elysium and Tartarus, or the State of the Final Retribution," the sixth panel of his massive "The Progress of Human Culture and Knowledge," painted for the Royal Society of Arts between 1777 and 1784. "It was my wish," Barry wrote, "to bring together in Elysium, those great and good men of all ages and nations, who were cultivators and benefactors of mankind; it forms a kind of apotheosis, or more properly a beatification of those useful qualities which were pursued through the whole work." Barry gathered some 125 men of genius into his heavenly pantheon, devoting the final panel to the arts and sciences. The theme, the promise of revelation in heaven of all knowledge that has remained obscure on earth, focuses on an archangel, with Newton on his right, explicating the workings of the solar system, with Bacon, Copernicus, Galileo, and Descartes in attendance.[28]

"Elysium" was the sixth and final canvas of James Barry's massive series of paintings, *The Progress of Human Knowledge and Culture*, executed for The Royal Society for Encouragement of Arts, Manufactures and Commerce (RSA) between 1777 and 1784. Newton is seen standing to the right of the Archangel, who explicates the workings of the solar system, revealing knowledge that has remained obscure on earth. – R.S.A., London, UK/ Bridgeman Art Library

Even these celebrations of Newton pale in comparison with the grandiose efforts by a group of French architects in the last quarter of the eighteenth century to commemorate the great Englishman in stone. The desire to do just that had been articulated as early as 1689 by the admiring Nicolas Fatio de Duillier, who had just arrived in England and was about to establish an intimate friendship with Newton. If "ever he got hold of an extra hundred thousand *écus*," Fatio wrote his brother, "he would erect statues and a monument" to honor Newton, "in order to let posterity know that during Newton's lifetime at least one man was capable of appreciating his worth."[29] Fatio, as noted above in chapter 2, had hoped to aggrandize his own stature through the glorification of Newton, and the same goal also informed the ideas of the revolutionary architect Etienne-Louis Boullée.

Boullée's design of a cenotaph for Newton is dated 1784. In his treatise on architecture, he discussed his vision, intended to match the depth of his veneration: "Sublime mind! Prodigious and profound genius! Divine Being! Newton! Deign to accept the homage of my feeble talents." Boullée's plan was as ingenious as it was grand: since Newton had defined the shape of the earth, Boullée conceived the idea of encasing Newton within his own discovery: "that is as it were to envelop you in your own self." What he conceptualized was a gigantic sphere by which he sought "to create the greatest of all effects, that of immensity." (If one calculates the ratio of the sphere to the minute figures placed in front of it, it appears that Boullée projected a sphere several hundred feet high!)

Intended to depict earth in its perfect shape at the moment of creation, before the motion ascribed to it by Newton had flattened it at the poles, the sphere was to be entirely empty – save for Newton's sarcophagus – in order not to distract attention from the grandeur and sublimity of the divine Newton (and, of course, the divine creator it also intended to commemorate). Internally, the perfect sphere was also designed to conjure up infinity in the soul of the viewer by its "continuous surface which has neither beginning nor end and the more we look at it (as in nature) the larger it appears." The dome was to be punctured to allow the rays of the sun to illuminate the interior during the day; a vast hanging "sepulchral lamp," in the shape of an armillary sphere, was to light up the structure at night. The cenotaph was envisaged as at once the ultimate memorial for Newton and a "temple of nature." The spectator "would become both the center and the circumference" of the anticipated ecstatic experience, "united with the tomb (Newton) and with an ever-expanding immensity (the cosmos)."[30]

BOULLEE'S CENOTAPH: MONUMENTAL GRANDEUR

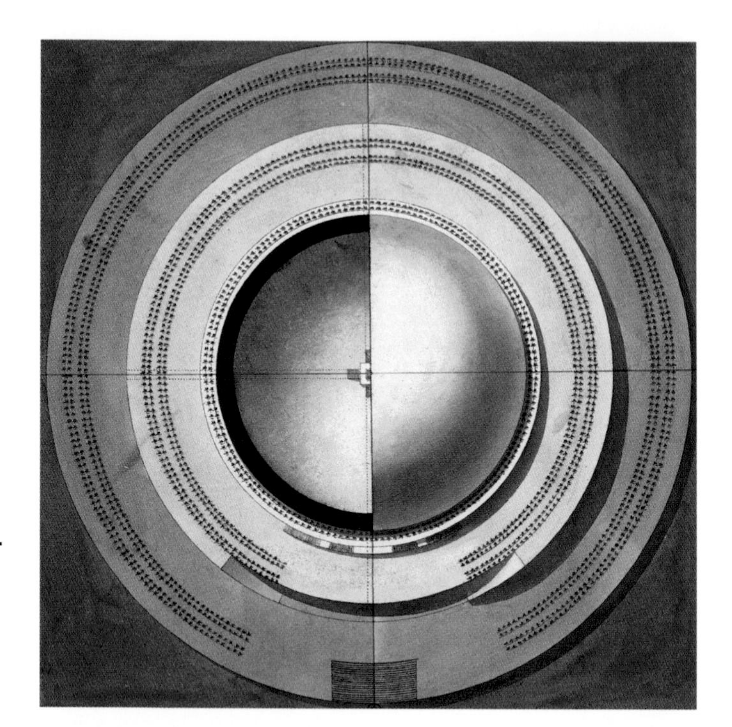

Etienne-Louis Boullée's design of a cenotaph for Newton (ca. 1785) was as ingenious as it was grand: since Newton had defined the shape of the earth, Boullée conceived the idea of encasing Newton within his own discovery. Boullée conceptualized a gigantic sphere that would impress upon the viewer the effects of infinity and immensity. Had it been built, the cenotaph would have been several hundred feet high. The six plates shown here present a schematic representation of the giant planetarium, and depict the cenotaph illuminated during both day and night. – Bibliothèque nationale de France

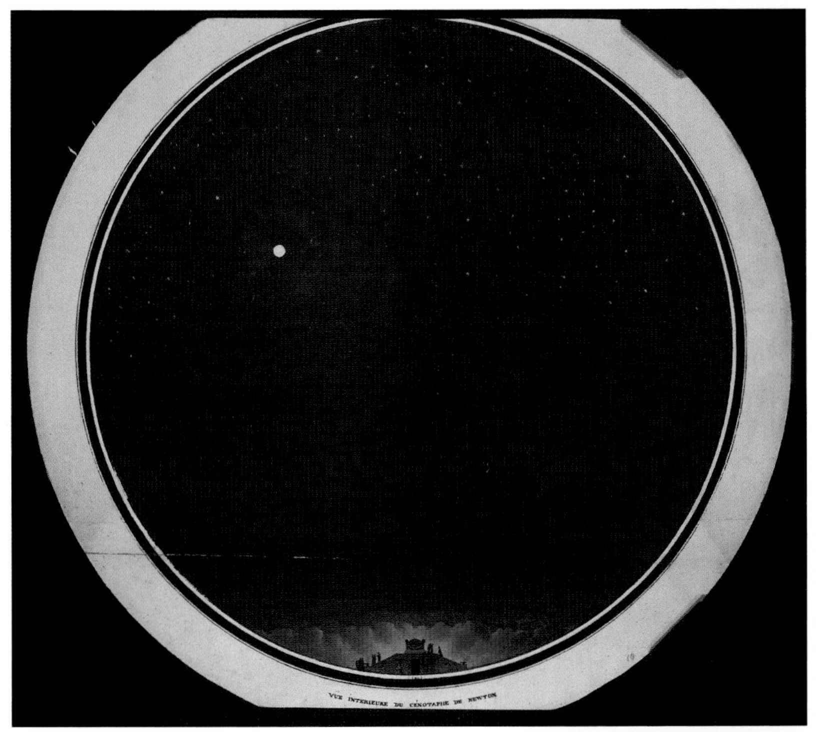
VUE INTERIEURE DU CÉNOTAPHE DE NEWTON

"Britannia Between Newton and Gravesande," Giovanni Martino dei Boni's homage to Newton and his great Dutch popularizer, was painted in 1787, most probably to commemorate the centenary of the *Principia*. - Arti Doria Pamphilj srl

Boullée was quite proud of his design, which, he boasted, conferred the sublime on him as well: "In order to obtain the natural tone and effect which are possible in this monument it was necessary to have recourse to all the magic of art and to paint with nature, i.e. to put nature to work; and I can say that this discovery belongs to me." It might be objected, he continued, that others had done similar things, but so, too, "apples fell before Newton," and "what was the result of it before this divine intelligence?"

Boullée's 1784 design was followed the next year by the announcement by the Academy of Architecture of a competition for a cenotaph for Newton, soliciting "an ingenious allegory, analogous to the writings of the great Newton" to serve as his monument. The winner was Pierre-Jules Delépine. In 1800, the Academy announced another cenotaph competition. Joseph-Jean-Pascal Gay won that competition with a design of a pyramid, each step of which was intended to depict an early astronomical system, all of them leading up to the true order of the universe discovered by Newton. The great man was to be seated "majestically on a throne" at the top of the pyramid, on his head an "aureola of seven rays," symbolizing the seven colors of his spectrum. Within the pyramid was imagined a large planetarium-like spherical room, on the blue ceiling of which the stars were to be drawn according to their true positions. A large globe would shed light. Newton's ashes would be contained in a shrine, while all his works would be engraved on marble tablets. As was the case with Boullée's cenotaph, Gay's design constituted both a memorial for Newton and a "temple to Nature" where Newton would live "through his works as well as in the memory of the living."[31]

As is often the case with grandiose schemes, none of these designs came even close to being realized. But the impulse behind them and other schemes to apotheosize Newton – to celebrate his genius and that of the new science he personified; to marvel at his perceived contribution to the glory of God's creation; and to articulate the wellsprings of one's own creativity by celebrating Newton's –

is characteristic of the era I term the "Newtonian moment." While the manifestations of this impulse are numerous, much has yet to be uncovered about them. We do not know, for example, the circumstances surrounding Giovanni Martino dei Boni's allegorical "Britannia Between Newton and Gravesand," undertaken in 1787, the first centenary of the *Principia*. So, too, the whereabouts of Robert-Guillaume Dardel's "Newton Discovers and Demonstrates the Truth," exhibited in the 1793 Paris Salon, remains a mystery, our knowledge of the statue being limited to a verbal description: Newton holding a prism in one hand (to honor his theory of colors) and a magnetized ring in the other (to denote his system of attraction). And just recently, knowledge has come to light that the celebrated neoclassical painter Giovanni Battista Piranesi was working in the late 1740s on his own rendition of the "Apotheosis of Sir Isaac Newton."[32]

As this chapter suggests, closer attention to the circumstances that led to execution of portraits and other visual artifacts should yield a rich harvest. Numerous examples exist of Enlightenment savants eager to depict kinship with Newton. Both Martin Folkes, President of the Royal Society, and the little-known English physician Benjamin Hoadley posed, like Benjamin Franklin, in proximity to a bust of Newton. John Bacon, an amateur scientific enthusiast, opted to arrange himself and his family against assorted scientific instruments, the likenesses of Newton and Sir Francis Bacon prominent in the background. Francesco Algarotti, who contributed so much to the popularization of Newtonian ideas, wished his memorial to say as much; his tomb – paid for by

Francesco Algarotti left detailed instructions regarding his tomb in the Campo Santo at Pisa, and Frederick the Great of Prussia – who paid for it – ensured that they were followed, including the wording of its simple inscription: "Algarotti, Emulator of Ovid and Newton's Disciple." – The Metropolitan Museum of Art

Arthur Devis's "John Bacon and His Family" (1742–43) depicts an archetypal Enlightenment gentleman savant in a family setting. Among the scientific instruments in the background are a telescope, an air pump, and a microscope. The four portraits in the middle distance include those of Sir Francis Bacon and Isaac Newton (upper left and bottom right, respectively), and on the far wall hangs a portrait of Edmond Halley. - Yale Center for British Art, Paul Mellon Collection, USA/Bridgeman Art Library

Frederick the Great of Prussia – thus carried the inscription "Newton's disciple."

By any standard, then, whether scientific or cultural, Newton's influence became ubiquitous. As the personification of science, he raised the stature of science and endowed his scientific contributions with such import that they soon became iconic images in contemporary art and literature. No matter that few, even in the scientific community, initially understood Newton "correctly" or embraced all his ideas. The revolution he effected gathered strength and speed as learned and popular treatises expounded or controverted his ideas, and other domains of knowledge increasingly coopted them. Eighteenth- and early nineteenth-century savants could adopt the real or perceived Newtonian vision, amend it, or elaborate on it. They could pick and choose parts from it; they could even reject it outright. What they could not do was to ignore it.

Even Johann Wolfgang von Goethe - one of the most virulent opponents of Newton's theory of colors – offered a supreme compliment when he chose to define his own greatness not by his literary achievements but by his distinction in understanding optics better than Newton. "What I have done as a poet," Goethe told a friend in old age, "I take no pride in it whatever. Excellent poets have lived at the same time with myself, poets more excellent have lived before me, and others will come after me. But that in my century I am the only person who knows the truth in the difficult science of colors – of that, I say, I am not a little proud, and here I have a consciousness of a superiority to many."[33] With this pronouncement, Goethe defined the essence of the "Newtonian moment."

By the middle of the eighteenth century, even the Jesuits were ready to apotheosize Newton, as is evident in the ceiling frescos painted for the Klementinum, the Mathematical Hall of the Jesuits' educational headquarters in Prague. One fresco (above) depicts a Newtonian universe, while in the other, the muse of astronomy is given Newton's facial likeness, thus suggesting his personification of the discipline. – Courtesy of the National Library of the Czech Republic

Globe du Soleil
Orbe de Mercure
Orbe de Venus
Orbe de la Terre
Orbe de Mars
Orbe de Jupiter
Orbe de Saturne
Orbe des Etoiles Fixes

NOTES

Throughout the text, contractions within seventeenth-century texts have been expanded while original spelling and punctuation have been retained.

Introduction

1 Immanuel Kant, *Critique of Pure Reason*, trans. Norman K. Smith (New York: St. Martin's Press, 1965), 19–25; I. Bernard Cohen, *Revolution in Science* (Cambridge, Mass.: Harvard University Press, 1985), 237–53; Robert Hahn, *Kant's Newtonian Revolution in Philosophy* (Carbondale: Southern Illinois University Press, 1988).

2 My understanding of "moment" is indebted to J.G.A. Pocock, *The Machiavellian Moment: Florentine Political Thought and the Atlantic Republican Tradition* (Princeton, N.J.: Princeton University Press, 1975).

Chapter 1

1 Richard S. Westfall, *Never at Rest: A Biography of Isaac Newton* (Cambridge: Cambridge University Press, 1980), 74, 191–92, et passim.

2 Richard S. Westfall, "Short-Writing and the State of Newton's Conscience, 1662," *Notes and Records of the Royal Society* 18 (1963): 10–16; Westfall, *Never at Rest*, 77–78.

3 Mordechai Feingold, "The Humanities" and "The Mathematical Science and the New Philosophies," in *The History of the University of Oxford*, ed. Nicholas Tyacke (Oxford: Oxford University Press, 1997), 211–448.

4 John Hall, *The Advancement of Learning* [1649], ed. A. K. Croston (Liverpool: Liverpool University Press, 1953), 26.

5 William Stukeley, *Memoirs of Sir Isaac Newton's Life* [1752], ed. Hastings White (London: Taylor and Francis, 1936), 53–54. For Barrow, see Mordechai Feingold, ed., *Before Newton: The Life and Times of Isaac Barrow* (Cambridge: Cambridge University Press, 1990).

6 Stukeley, *Memoirs of Sir Isaac Newton's Life*, 54.

7 D. T. Whiteside, "Newton the Mathematician," in *Contemporary Newtonian Research*, ed. Zev Bechler (Dordrecht and Boston: D. Reidel Publishing Company, 1982), 109–27, at 110–11; Stukeley, *Memoirs of Sir Isaac Newton's Life*, 54–55.

8 A. Rupert Hall, "Sir Isaac Newton's Note-Book, 1661–65," *Cambridge Historical Journal* 9 (1948): 239–50; William A. Wallace, "Newton's Early Writings: Beginnings and New Directions," in *Newton and the New Directions in Science*, ed. G. V. Coyne, S. M. Heller, and J. Zycinski (Vatican City: Vatican Observatory, 1988), 23–44; Maurizio Mamiani, "To Twist the Meaning: Newton's *Regulae Philosophandi* Revisited," in *Isaac Newton's Natural Philosophy*, ed. Jed Z. Buchwald and I. Bernard Cohen (Cambridge, Mass.: MIT Press, 2001), 3–14; Maurizio Mamiani, "Newton on Prophecy and the Apocalypse," in *The Cambridge Companion to Newton*, ed. I. Bernard Cohen and George E. Smith (Cambridge: Cambridge University Press, 2002), 387–408.

9 William Watts, "To the Venerable Artists and Younger Students in Divinity, in the Famous University of Cambridge," in *The Strange and Dangerous Voyage of Captaine Thomas James* (London, 1633), sig. S; J. E. McGuire and Martin Tamny, *Certain Philosophical Questions: Newton's Trinity Notebook* (Cambridge: Cambridge University Press, 1983), 337.

10 McGuire and Tamny, *Certain Philosophical Questions*, 482–83; *The Correspondence of Isaac Newton*, ed. H. W. Turnbull, J. F. Scott, A. R. Hall, and Laura Tilling, 7 vols. (Cambridge: Cambridge University Press, 1959–77), 3: 153; Rob Iliffe, "'That Puzleing Problem': Isaac

Newton and the Political Physiology of Self," *Medical History* 39 (1995): 433–58, at 439–45.

11 J. E. McGuire and Martin Tamny, "Newton's Astronomical Apprenticeship: Notes of 1664/5," *Isis* 76 (1985): 349–65.

12 *The Mathematical Papers of Isaac Newton*, ed. D. T. Whiteside, 8 vols. (Cambridge: Cambridge University Press, 1967–81), 1: 5–6, 17–18.

13 Newton, *Mathematical Papers*, 1: 7–8.

14 D. T. Whiteside, "Newton's Marvellous Year: 1666 and All That," *Notes and Records of the Royal Society* 21 (1966): 32–41, at 34.

15 Mordechai Feingold, "Newton, Leibniz, and Barrow Too: An Attempt at a Reinterpretation," *Isis* 84 (1993): 310–38.

16 Cambridge University Library, MS Add 3968, fol. 85; Whiteside, "Newton's Marvellous Year," 32; Westfall, *Never at Rest*, 106–38.

17 Cambridge University Library, MS Add 3968.41, fol. 85r; Westfall, *Never at Rest*, 143.

18 Westfall, *Never at Rest*, 154–55; Stukeley, *Memoirs of Sir Isaac Newton's Life*, 20; J. Bruce Brackenridge and Michael Nauenberg, "Curvature in Newton's Dynamics," in *The Cambridge Companion to Newton*, 85–137.

19 Newton, *Correspondence*, 1: 3.

20 *Unpublished Scientific Papers of Isaac Newton*, ed. A. Rupert Hall and Marie Boas Hall (Cambridge: Cambridge University Press, 1978), 403; *The Optical Papers of Isaac Newton*, ed. Alan Shapiro (Cambridge: Cambridge University Press, 1984), 9; Alan E. Shapiro, *Fits, Passions, and Paroxysms: Physics, Method, and Chemistry and Newton's Theories of Colored Bodies and Fits of Easy Reflection* (Cambridge: Cambridge University Press, 1993), 102–5.

21 Newton, *Correspondence*, 1: 82, 92–102, 107-8.

22 Thomas Birch, *The History of the Royal Society of London*, 4 vols. (repr. Brussels: Culture et Civilisation, 1967), 3: 9–10; *Philosophical Transactions* 6 (1672): 3075–87.

23 *The Correspondence of Henry Oldenburg*, ed. A. Rupert Hall and Mary B. Hall, 13 vols. (Madison: University of Wisconsin Press; London: Mansell; London: Taylor and Francis, 1965–86), 9: 249; Newton, *Correspondence*, 1: 96–97.

24 Newton, *Correspondence*, 1: 110–14.

25 *The Diary of Robert Hooke ... 1672–1680*, ed. Henry W. Robinson and Walter Adams (repr. London: Wykeham Publications, 1968), 213; Newton, *Correspondence*, 1: 412–13.

26 Newton, *Correspondence*, 1: 356.

27 Ibid., 2: 297–98.

28 Ibid., 2: 300–308.

29 Ibid., 2: 309–10, 312–13.

30 Ibid., 2: 440–47.

31 Ibid., 2: 435–40.

32 Isaac Newton, *The Principia*, trans. I. Bernard Cohen and Anne Whitman (Berkeley: University of California Press, 1999), 793.

33 King's College, Cambridge, MS Keynes 130.

34 Newton, *Correspondence*, 1: 416–17; Frank E. Manuel, *A Portrait of Isaac Newton* (Cambridge, Mass.: Harvard University Press, 1968), 145–46.

35 *Memoirs of the Life and Writings of Mr. William Whiston ... Written by Himself*, 2 vols. (London, 1749), 1: 35–36; King's College, Cambridge, MS Keynes 160.6; Newton, *Mathematical Papers*, 4: 336–45.

36 Algebra, Newton told David Gregory in 1708, "is the Analysis of the Bunglers in Mathematicks." For his part, Henry Pemberton heard the great man lament that he began his mathematical studies by "applying himself to the works of *Des Cartes* and other algebraic writers, before he had considered the elements of *Euclide* with that attention, which so excellent a writer deserves." Newton, *Mathematical Papers*, 4: 275–77; Stukeley, *Memoirs of Sir Isaac Newton's Life*, 42; Henry Pemberton, *A View of Sir Isaac Newton's Philosophy* (London, 1728), sig. a; Niccolò Guicciardini, *Reading the Principia: The Debate on Newton's Mathematical Methods for Natural Philosophy from 1687 to 1736* (Cambridge: Cambridge University Press, 1999), 27–32, 101–6.

37 Newton, *Principia*, 46, 165, 416, 424.

38 That the document in its present form is of late composition, albeit incorporating earlier material, can be deduced from its invocation of the concept of inertia, a term not used by Newton before the mid-1680s.

39 Bodleian Library, MS Smith 8, fol. 147; Roger North, *Notes of Me: The Autobiography of Roger North*, ed. Peter Millard (Toronto: University of Toronto Press, 2000), 92. *De gravitatione* is

printed and translated in *Unpublished Scientific Papers of Isaac Newton*, 89–156.

40 Newton, *Principia*, 588–89.

Chapter 2

1 Westfall, *Never at Rest*, 403.
2 Stephen P. Rigaud, *Historical Essay on the First Publication of Sir Isaac Newton's Principia* (Oxford, 1838), 35.
3 Edmond Halley, *Correspondence and Papers of Edmond Halley*, ed. Eugene F. MacPike (New York: Arno Press, 1975), 80.
4 Ibid., 63–64, 68–69.
5 *Philosophical Transactions* 16 (1687): 291–97.
6 I. Bernard Cohen, *Introduction to Newton's 'Principia'* (Cambridge, Mass.: Harvard University Press, 1971), 145–52, 156–57; James L. Axtell, "Locke's Review of the *Principia*," *Notes and Records of the Royal Society* 20 (1965): 152–61; I. Bernard Cohen, *The Newtonian Revolution* (Cambridge, Mass.: Harvard University Press, 1980), 96–98.
7 *Isaac Newton: Eighteenth-century Perspectives*, ed. A. Rupert Hall (Oxford: Oxford University Press, 1998), 61; Jean-Baptiste Biot, *Life of Sir Isaac Newton* (London, 1841), 23.
8 Cohen, *Introduction to Newton's 'Principia,'* 299; Newton, *Correspondence*, 2: 485, 493.
9 Newton, *Correspondence*, 2: 501 and n. 2, 3: 7; Westfall, *Never at Rest*, 471.
10 Robert S. Westman, "Huygens and the Problem of Cartesianism," in *Studies on Christiaan Huygens*, ed. Henk J. M. Bos (Lisse: Swets & Zeitlinger, 1980), 83–103, at 99.
11 Alexandre Koyré, *Newtonian Studies* (London: Chapman & Hall, 1965), 115–24; H.A.M. Snelders, "Christiaan Huygens and Newton's Theory of Gravitation," *Notes and Records of the Royal Society* 43 (1989): 209–22; Alan Shapiro, "Huygens' 'Traité de la Lumière' and Newton's 'Opticks': Pursuing and Eschewing Hypotheses," *Notes and Records of the Royal Society* 43 (1989): 223–47.
12 Charles A. Domson, *Nicolas Fatio de Duillier and the Prophets of London* (New York: Arno Press, 1981), 32–33.
13 Newton, *Correspondence*, 3: 69, cited in A. Rupert Hall, *Isaac Newton, Adventurer in Thought* (Cambridge: Cambridge University Press, 1996), 234.
14 Nicolas Fatio de Duillier, "De la cause de la pesanteur," ed. Bernard Gagnebin, *Notes and Records of the Royal Society* 6 (1949): 105–60, at 110, 115–17; Cohen, *Introduction to Newton's 'Principia,'* 180.
15 Newton, *Correspondence*, 3: 258, 286–87.
16 Newton, *Principia*, 943.
17 Newton, *Correspondence*, 3: 4–5; Domenico Bertoloni Meli, *Equivalence and Priority: Newton Versus Leibniz* (Oxford: Clarendon Press, 1993), 7–8.
18 Newton, *Correspondence*, 3: 372–73, cited in Manuel, *A Portrait of Isaac Newton*, 222; Westfall, *Never at Rest*, 721.
19 Newton, *Correspondence*, 3: 258.
20 Rigaud, *Historical Essay*, 96; A. Rupert Hall, *Philosophers at War: The Quarrel Between Newton and Leibniz* (Cambridge: Cambridge University Press, 1980), 100.
21 Meli, *Equivalence and Priority*, 11.
22 Newton, *Mathematical Papers*, 8: 72–75.
23 Hall, *Philosophers at War*, 105–6; Westfall, *Never at Rest*, 581–83.
24 Hall, *Philosophers at War*, 119–26; Nicolas Fatio de Duillier, *Lineae brevissimi descensus investigatio geometrica duplex* (London, 1699); Newton, *Correspondence*, 5: 98 n. 3.
25 Isaac Newton, *Opticks*, ed. Edmund Whittaker and I. Bernard Cohen (New York: Dover Publications, 1952), cxxi.
26 Alan E. Shapiro, "The Gradual Acceptance of Newton's Theory of Light and Color, 1672–1727," *Perspectives on Science* 4 (1996): 59–140, at 102–4; Hall, *Philosophers at War*, 138–40.
27 Hall, *Philosophers at War*, 159–61; Meli, *Equivalence and Priority*, 187, 206–7.
28 John Locke, *An Essay Concerning Human Understanding*, ed. Peter H. Nidditch (Oxford: Clarendon Press, 1975), 9–10, 599.
29 Nicholas Jolley, *Leibniz and Locke* (Oxford: Oxford University Press, 1984), 36–37, 54–55; Hans Aarsleff, *From Locke to Saussure* (Minneapolis: University of Minnesota Press, 1982), 51; *The Leibniz-Clarke Correspondence*, ed. H. G. Alexander (Manchester: Manchester University Press, 1956), II; G. W. Leibniz, *The Monadology and Other Philosophical Writings*

(Oxford: Oxford University Press, 1968), 388.

30 John Keill, "The Laws of Centripetal Force," *Philosophical Transactions* 26 (1708, published 1710): 185; Hall, *Philosophers at War*, 144–45.

31 Newton, *Correspondence*, 5: 97.

32 During that same period, John Flamsteed, the first Astronomer Royal, found himself the victim of Newton's covert effort to force Flamsteed to publish his Star Catalogue. He, too, refused to believe for a long time that it was Newton himself, and not his lackeys (Edmond Halley in particular), who had sought to undermine his position and wrest the catalogue from him.

33 Newton, *Correspondence*, 5: 142–49; Hall, *Philosophers at War*, 169–75.

34 Newton, *Correspondence*, 5: 143, 145, 148.

35 Ibid., 2: 115, 134; Westfall, *Never at Rest*, 265.

36 Newton, *Correspondence*, 5: 207–8.

37 Hall, *Philosophers at War*, 193; Newton, *Correspondence*, 6: 18–19.

38 William Whiston, *Historical Memoirs of the Life of Dr. Samuel Clarke* (London, 1730), 132.

39 Newton, *Correspondence*, 5: 143, 203, 298–99.

40 Ibid., 5: 348–50, 400; Hall, *Philosophers at War*, 193.

Chapter 3

1 Newton, *Correspondence*, 5: 116 n. 6, 299–300, 389–90, 392; *Die philosophischen Schriften von Gottfried Wilhelm Leibniz*, ed. C. I. Gerhardt, 7 vols. (repr. Hildesheim and New York: G. Olms, 1978), 3: 519.

2 Newton, *Principia*, 385–86.

3 Ibid., 198–200, 794–96, 939–44.

4 Westfall, *Never at Rest*, 587; Newton, *Correspondence*, 6: 59.

5 Joseph Bertrand, *L'Académie des Sciences et les Academiciens de 1666 à 1793* (Paris, 1869), 37–38. For Halley's précis of the book, specially prepared for James II, see Newton, *Correspondence*, 2: 483; *Philosophical Transactions* 19 (1695–97): 445–57.

6 Newton, *Correspondence*, 6: 42.

7 Newton, *Principia*, 382 (italics in the original).

8 Paolo Mancosu, "The Metaphysics of the Calculus: A Foundational Debate in the Paris Academy of Sciences, 1700–1706," *Historia Mathematica* 16 (1989): 224–48; Mancosu, *Philosophy of Mathematics and Mathematical Practice in the Seventeenth Century* (Oxford: Oxford University Press, 1996), 165–77; John B. Shank, "Before Voltaire: Newtonianism and the Origins of the Enlightenment in France, 1687–1734," Ph.D. thesis (Stanford University, 2000), 215–97.

9 M. de Fontenelle (Bernard le Bovier), *Entretiens sur la pluralité des mondes*, ed. Robert Shackleton (Oxford: Clarendon Press, 1955), 21–24.

10 Shank, "Before Voltaire," 304–5.

11 *The Letters of Joseph Addison*, ed. Walter Graham (Oxford: Clarendon Press, 1941), 25.

12 Henry Guerlac, *Newton on the Continent* (Ithaca, N.Y.: Cornell University Press, 1981), 107–11; Shapiro, "The Gradual Acceptance of Newton's Theory of Light and Color, 1672–1727," 97–100. See also Paul Mouy, "Malebranche et Newton," *Revue de Métaphysique et de Morale* 45 (1938): 411–35.

13 Malebranche certainly attended the ten sessions (August 1706–June 1707) during which Geoffrey read his account, which was based on the English version forwarded to him by Sir Hans Sloane. Among the other attendees at Geoffroy's reading was Jean Truchet (Père Sébastien) who would become, a decade later, instrumental in another determined effort to confirm Newton's experiments. I. Bernard Cohen, "Isaac Newton, Hans Sloane and the Académie Royale des Sciences," in *Mélanges Alexandre Koyré*, 2 vols. (Paris: Hermann, 1964), 2: 61–116, at 105–16; Guerlac, *Newton on the Continent*, 100–103.

14 Newton, *Correspondence*, 7: 114, 157–58; Guerlac, *Newton on the Continent*, 139–44.

15 Newton, *Correspondence*, 6: 188–89; John L. Greenberg, *The Problem of the Earth's Shape from Newton to Clairaut* (Cambridge: Cambridge University Press, 1995), 76; A. Rupert Hall, "Newton in France," *History of Science* 13 (1975): 233–50, at 239–41.

16 Greenberg, *The Problem of the Earth's Shape*, 115.

17 Joseph Spence, *Observations, Anecdotes, and Characters of Books and Men*, ed. James M. Osborn, 2 vols. (Oxford: Clarendon Press, 1966), 1: 461.

18 *Isaac Newton: Eighteenth-century Perspectives*,

ed. A. Rupert Hall (Oxford: Oxford University Press, 1998), 61, 65.

19 Newton, *Principia*, 529, 731; Newton, *Mathematical Papers*, 6: 349 n. 209, 466–67 n. 25.

20 Guicciardini, *Reading the Principia*, 202; Greenberg, *The Problem of the Earth's Shape*, 12.

21 Greenberg, *The Problem of the Earth's Shape*, 13.

22 Alan E. Shapiro, "Huygens' 'Traité de la lumière' and Newton's 'Opticks': Pursuing and Eschewing Hypotheses," *Notes and Records of the Royal Society* 43 (1989): 223–47, at 226.

23 Patricia Reif, "The Textbook Tradition in Natural Philosophy, 1600–1650," *Journal of the History of Ideas* 30 (1969): 17–32, at 29.

24 *Memoirs of the Life and Writings of Mr. William Whiston*, 1: 36; Whiston, *Historical Memoirs of the Life of Dr. Samuel Clarke*, 6; Michael A. Hoskin, "'Mining All Within': Clarke's Notes to Rohault's *Traité de physique*," *The Thomist* 24 (1961): 353–63.

25 Samuel Golden, *Jean Le Clerc* (New York: Twayne, 1972), 86.

26 Jean Le Clerc, "Eloge de feu Mr. de Volder," *Bibliothèque Choisie* 18 (1709): 346–401; Edward G. Ruestow, *Physics at 17th- and 18th-century Leiden* (The Hague: Martinus Nijhoff, 1973), 89–112; C. de Pater, "Experimental Physics," in *Leiden University in the Seventeenth Century: An Exchange of Learning*, ed. Th. H. Lunsingh Scheurleer and G.H.M. Posthumus Meyjes (Leiden: Brill, 1975), 309–27, at 314–18; Wim N. A. Klever, "Burchard de Volder (1643–1709). A Crypto-Spinozist on a Leiden Cathedra," *Lias* 15 (1988): 191–241.

27 Archibald Pitcairne, *The Philosophical and Mathematical Elements of Physick*, trans. John Quincy, 2nd ed. (London, 1745), xvii–xxviii; Anita Guerrini, "Archibald Pitcairne and Newtonian Medicine," *Medical History* 31 (1987): 70–83.

28 *Boerhaave's Orations*, trans. E. Kegel-Brinkgreve and A. M. Luyendijk-Elshout (Leiden: Brill, 1983), 115, 132, 155–79.

29 Ibid., 212.

30 Julien Offray de la Mettrie, "Vie de M. Herman Boerhaave," in *Institutions de médicine de M. Herman Boerhaave* (1740), cited in G. A. Lindeboom, *Herman Boerhaave: The Man and His Work* (London: Methuen, 1968), 270.

31 Ruestow, *Physics at 17th- and 18th-century Leiden*, 113–39; A. Rupert Hall, "Further Newton Correspondence," *Notes and Records of the Royal Society* 37 (1982): 7–34, at 26; G. B. Gori, *La fondazione dell' esperienza in 's Gravesande* (Florence: La nuova Italia, 1972).

32 Peter de Clerq, *At the Sign of the Oriental Lamp: The Musschenbroek Workshop in Leiden, 1660–1750* (Rotterdam: Erasmus Publishing, 1997), 48.

33 C. de Pater, "Petrus van Musschenbroek (1692–1761): A Dutch Newtonian," *Janus* 64 (1977): 77–87.

34 Pietro Redondi, *Galileo: Heretic* (Princeton, N.J.: Princeton University Press, 1987), 307–8; Susana Gomez Lopez, *Le passioni degli atomi. Montanari e Rossetti: una polemica tra galileiani* (Florence: Leo S. Olschki, 1997); Luciano Osbat, *L'Inquisizione a Napoli. Il processo agli ateisti. 1688-1697* (Rome: Edizioni di storia e letteratura, 1974).

35 Michael Segre, *In the Wake of Galileo* (New Brunswick, N.J.: Rutgers University Press, 1991), 141; Marta Cavazza, *Settecento inquieto: Alle origini dell'Istituto delle Scienze di Bologna* (Bologna: Il Mulino, 1990), 215, cited in J. L. Heilbron, *The Sun in the Church: Cathedrals as Solar Observatories* (Cambridge, Mass.: Harvard University Press, 1999), 203.

36 Claudio Constantini, *Baliani e i Gesuiti* (Florence: Giunti, 1969), 108; Mordechai Feingold, "Jesuits: Savants," in *Rethinking Jesuit Science*, ed. Mordechai Feingold (Cambridge, Mass.: MIT Press, 2002), 1–45.

37 Ugo Baldini, "Teoria Boscovichiana, Newtonismo, Eliocentrismo: Dibattiti nel Collegio Romano e nella Congregazione dell'Indice a metà Settecento," in *Saggi sulla Cultura della Compagnia di Gesù (secoli XVI-XVIII)* (Padua: CLEUP, 2000), 283–95.

38 Vincenzo Ferrone, *The Intellectual Roots of the Italian Enlightenment: Newtonian Science, Religion, and Politics in the Early Eighteenth Century*, trans. Sue Brotherton (Atlantic Highlands, N.J.: Humanities Press, 1995), 279–80 n. 21; Germano Paoli, "Boscovich and Enlightenment," in *Bicentennial Commemoration of R. G. Boscovich*, ed. M. Bossi and P.

Tucci (Milan: Edizioni Unicopli, 1988), 232–33.

39 A. Fabroni, *Lettere inedite d'uomini illustri*, 2 vols. (Florence, 1773–75), 2: 161, cited in Brendan Dooley, "The Communication Revolution in Italian Science," *History of Science* 33 (1995): 469–96, at 470.

40 Cesare Beccaria, *On Crimes and Punishments and Other Writings*, ed. Richard Bellamy, Richard Davies, and Virginia Cox (Cambridge: Cambridge University Press, 1995), 121.

41 Heilbron, *The Sun in the Church*, 197; Ferrone, *The Intellectual Roots of the Italian Enlightenment*, 292 n. 4; John L. Russell, "Catholic Astronomers and the Copernican System After the Condemnation of Galileo," *Annals of Science* 46 (1989): 365–86, at 383.

42 Gino Tellini, "Tra corrispondenti di Francesco Redi: lettere inedite di Geminiano Montanari, Francesco d'Andrea, e Paolo Boccone," *Filologia e Critica* 1 (1976): 418; Dooley, "The Communication Revolution in Italian Science," 125; Ferrone, *The Intellectual Roots of the Italian Enlightenment*, 35, 203.

43 Segre, *In the Wake of Galileo*, 138; Paolo Galluzzi, "Galileo contro Copernico," *Annali dell' Istituto e Museo di Storia della Scienza di Firenze* 2 (1977): 87–148.

44 Eric Cochrane, *Florence in the Forgotten Centuries 1527–1800* (Chicago: University of Chicago Press, 1974), 253; Brendan Dooley, "The Giornale de' Letterati d'Italia (1710–40): Journalism and 'Modern' Culture in the Early Eighteenth Century Veneto," *Studi Veneziani*, n.s. 6 (1982): 229–70, at 245.

45 Ferrone, *The Intellectual Roots of the Italian Enlightenment*, 41; Domenico Bertoloni Meli, "Leibniz on the Censorship of the Copernican System," *Studia Leibnitiana* 20 (1988): 19–42; Manuel, *A Portrait of Isaac Newton*, 267.

46 Ferrone, *The Intellectual Roots of the Italian Enlightenment*, 15; A. Rupert Hall, "Further Newton Correspondence," *Notes and Records of the Royal Society* 37 (1982): 7–34, at 17–18; *The Life, Unpublished Letters, and Philosophical Regimen of Anthony, Earl of Shaftesbury*, ed. Benjamin Rand (London: Swan Sonnenschein, 1900), 479, 497.

47 Paolo Casini, *Newton e la cosciena europea* (Bologna: Il Mulino, 1983), 191; Paola Zambelli, "Antonio Genovesi and Eighteenth-century Empiricism in Italy," *Journal of the History of Philosophy* 16 (1978): 195–208, at 196–97 n. 9; *Natura e humanitas del Giovane Vico*, ed. Maria Donzelli (Naples: Istituto Italiano per Gli Studi Storici, 1970), 164–88; R. Gatto, G. Gerla, and F. Palladino, "Lettere di Giacinto de Cristofaro a Bernard Fontenelle e a Celestino Galiani," *Annali dell'Istituto e Museo di Storia della Scienza di Firenze* 9 (1984): 67–93, at 80–83.

48 *Opere del conte Jacopo Riccati, nobile trevigiano*, 4 vols. (Lucca, 1761–65), 4: vii; Hall, "Further Newton Correspondence," 15–16; Newton, *Correspondence*, 7: 434; Silvia Mazzone and Clara Silvia Roero, *Jacob Hermann and the Diffusion of the Leibnizian Calculus in Italy* (Florence: Leo S. Olschki, 1997), 190; Cesare S. Maffioli, *Out of Galileo: The Science of Waters 1628–1718* (Rotterdam: Erasmus Publishing, 1994), 252–72.

49 Brendan Dooley, "Science Teaching as a Career at Padua in the Early Eighteenth Century: The Case of Giovanni Poleni," *History of Universities* 4 (1984): 115–51; Maffioli, *Out of Galileo*, 327–34; Mazzone and Roero, *Jacob Hermann and the Diffusion of the Leibnizian Calculus*, 142–46.

50 Maffioli, *Out of Galileo*, 339–40, 350; Ferrone, *The Intellectual Roots of the Italian Enlightenment*, 27.

51 Newton, *Correspondence*, 6: 279–80, 331.

52 Mazzone and Roero, *Jacob Hermann and the Diffusion of the Leibnizian Calculus*, 322 et passim; Hall, "Further Newton Correspondence," 22–23.

53 Stephen P. Rigaud, *Correspondence of Scientific Men of the Seventeenth Century*, 2 vols. (Oxford, 1841), 2: 279–80; I. Bernard Cohen, *The Newtonian Revolution*, 143–44; Mazzone and Roero, *Jacob Hermann and the Diffusion of the Leibnizian Calculus*, 75 n. 137.

54 Ferrone, *The Intellectual Roots of the Italian Enlightenment*, 26.

55 Ibid., 10–13, 18–21, 124–30; Mazzone and Roero, *Jacob Hermann and the Diffusion of the Leibnizian Calculus*, 294–308.

56 Vincenzo Ferrone, "Celestino Galiani e la diffusione del newtonianesimo: Appunti e documenti per una storia della cultura scientifica del primo

Settecento," *Giornale Critico della Filosofia Italiana* 2 (1982): 1–33; Ferrone, *The Intellectual Roots of the Italian Enlightenment*, 28–31.

57 Ferrone, *The Intellectual Roots of the Italian Enlightenment*, 129, 305 n. 25, 333 n. 17.

58 Ibid., 338 n. 33, 41.

59 Marta Cavazza, "Bologna and the Royal Society in the Seventeenth Century," *Notes and Records of the Royal Society* 35 (1980): 118.

60 C. de Pater, "The Textbooks of 's Gravesande and Van Musschenbroek in Italy," in *Italian Scientists in the Low Countries in the XVII and XVIIIth Centuries*, ed. C. S. Maffioli and L. C. Palm (Amsterdam: Rodopi, 1989), 231–41, at 231; Ferrone, *The Intellectual Roots of the Italian Enlightenment*, 97–99.

61 Roger Joseph Boscovich, *A Theory of Natural Philosophy* (Cambridge, Mass.: MIT Press, 1966), 6–7; Juan Casanovas, "Boscovich as an Astronomer," in *Bicentennial Commemoration of R. G. Boscovich*, 57–70, at 62–63.

62 Maffioli, *Out of Galileo*, 21 and n. 10; Paula Findlen, "Translating the New Science: Women and the Circulation of Knowledge in Enlightenment Italy," *Configurations* 2 (1995): 167–206, at 191.

63 Zambelli, "Antonio Genovesi and Eighteenth-century Empiricism in Italy," 199–200.

64 Ferrone, *The Intellectual Roots of the Italian Enlightenment*, 111–15.

65 Casini, *Newton e la cosciena europea*, 202–7; Shapiro, "The Gradual Acceptance of Newton's Theory of Light and Color, 1672–1727," 125–30.

Chapter 4

1 André Michel Rousseau, *L'Angleterre et Voltaire*, 2 vols. (Oxford: Voltaire Foundation, 1976), 1: 113; Theodore Besterman, *Voltaire*, 3rd ed. (Chicago: University of Chicago Press, 1976), 39.

2 Voltaire, *Correspondence*, in *The Complete Works of Voltaire*, ed. Theodore Besterman (Geneva: Institut et Musée Voltaire, and Oxford: Voltaire Foundation, 1968–), 85: 203–4 (D190). As Bolingbroke insisted, "it is not a question of opinion, or of probability, that nature does not act as it is made to act."

3 Alfred O. Aldridge, *Voltaire and the Century of Light* (Princeton, N.J.: Princeton University Press, 1975), 44.

4 John Morley, *Voltaire* (London: Macmillan and Co., 1903), 58. And, Morley added, "what was not any less important, he had become alive to the central truth of the social destination of all art and knowledge."

5 Voltaire, *Correspondence*, 85: 303 (D299), 309–10 (D303), 333 (D330).

6 Voltaire, *Letters on England*, trans. Leonard Tancock (Harmondsworth: Penguin Books, 1984), 69, 112.

7 S. H. Barber, "Voltaire and Samuel Clarke," *Studies on Voltaire and the Eighteenth Century* 179 (1979): 47–61.

8 Voltaire, *Correspondence*, 85: 320 (D315).

9 Voltaire, *Notebooks*, ed. Theodore Besterman, in *The Complete Works of Voltaire*, 81: 76, 83, 86–87, 92–93.

10 Voltaire, *The English Essays of 1727*, ed. Penelope Brading, in *The Complete Works of Voltaire* (Oxford: Voltaire Foundation, 1996), 3b: 6–7, 372–73; Voltaire, *La Henriade*, ed. O. R. Taylor, in *The Complete Works of Voltaire* (Geneva: Institut et Musée Voltaire, 1970), 2: 328–29, 511–13.

11 Voltaire, *Letters on England*, 63, 68, 71–72.

12 Voltaire, *Correspondence*, 86: 389 (D653).

13 Ibid., 86: 342–43 (D617); Horst-Heino von Borzeszkowski and Renate Wahsner, *Voltaire's Newtonianism: A Bridge from English Empiricism to Cartesian Rationalism and Its Implications for the Concept of Mechanics in German Idealism* (Berlin: Max-Planck-Inst. für Wissenschaftsgeschichte, 2000), 23 n. 9.

14 Voltaire, *Correspondence*, 86: 266 (D550).

15 Ibid., 87: 132 (D863), 134 (D865).

16 Ibid., 88: 196 (D1255).

17 Robert L. Walters, "The Allegorical Engravings in the Ledet-Desbordes Edition of the *Eléments de la philosophie de Newton*," in *Voltaire and His World*, ed. R. J. Howells et al. (Oxford: Voltaire Foundation, 1985), 27–49, at 44–46.

18 Voltaire, *Eléments de la philosophie de Newton*, ed. Robert L. Walters and W. H. Barber (Oxford: Voltaire Foundation, 1992), 84–85; Besterman, *Voltaire*, 201.

19 Voltaire, *Correspondence*, 91: 14 (D2088); 93: 30–31 (D2890); 117: 449 (D15,140).

20 Ibid., 86: 251 (D537); 88: 66 (D1154). See Paolo Casini, "Briarée en miniature: Voltaire et Newton," *Studies in Voltaire and the Eighteenth Century* 179 (1979): 63–79.
21 Voltaire, *Correspondence*, 89: 244 (D1578).
22 Ibid., 88: 89 (D1174).
23 Ibid., 90: 283 (D1936), cited in Jean Dhombres, "Books: Reshaping Science," in *Revolution in Print: The Press in France, 1775–1800*, ed. Robert Darnton and Daniel Roche (Berkeley: University of California Press, 1989), 177–202, at 187.
24 Aldridge, *Voltaire*, 108–9.
25 *Histoire de l'esprit humain* 4 (1765–68): 311–12, cited in Andreas Kleinert, "La vulgarisation de la physique au siècle des lumières," *Francia* 10 (1982): 303–12, at 307.
26 *Historical and Literary Memoirs and Anecdotes, Selected from the Correspondence of Baron de Grimm and Diderot with the Duke of Saxe-Gotha Between the Years 1770 and 1790*, 2 vols. (London, 1814), 1: 108–9; Isaac Newton, *Principes mathématiques de la philosophie naturelle*, trans. Marquise du Châtelet, 2 vols. (Paris, 1759), 1: viii.
27 Besterman, *Voltaire*, 202–3.
28 Voltaire, *Correspondence*, 89: 28 (D1448); Voltaire, *Eléments*, 83–84.
29 Charles B. Paul, *Science and Immortality: The Eloges of the Paris Academy of Sciences (1699–1791)* (Berkeley: University of California Press, 1980), 30.
30 Voltaire, *Eléments*, 90; Voltaire, *Correspondence*, 89: 268–70 (D1600); Ira O. Wade, *The Intellectual Development of Voltaire* (Princeton, N.J.: Princeton University Press, 1969), 415–16.
31 Voltaire, *Correspondence*, 89: 307–20 (D1622), 286–87 (D1611); Voltaire, *Eléments*, 89, 125, 671, 698–718, 733.
32 *Voltariana, ou éloges amphigouriques de Fr. Marie Arrouet* (Paris, 1748), 59.
33 Henry Guerlac, *Newton on the Continent* (Ithaca, N.Y.: Cornell University Press, 1981), 131.
34 Grimm, *Historical and Literary Memoirs and Anecdotes*, 1: 108–9.
35 *Turgot on Progress, Sociology and Economics*, trans. Ronald L. Meek (Cambridge: Cambridge University Press, 1973), 58–59, 94–96.
36 Jean Le Rond d'Alembert, *Preliminary Discourse to the Encyclopedia of Diderot*, trans. Richard N. Schwab (Indianapolis: Bobbs-Merrill, 1976), 78, 81.
37 Marquis d'Argens, *Chinese Letters* (New York: Garland Publishing, 1974), 113–14.
38 Denis Diderot, *The Indiscreet Jewels* [1748], trans. Sophie Hawkes (New York: Marsilio, 1993), 28–29.
39 Edward Gibbon, *Memoirs of My Life*, ed. Georges Bonnard (New York: Funk & Wagnalls, 1966), 125–26.
40 Antoine Guénard, *Discours sur l'esprit philosophique* (Paris, 1755), in A. Cahour, *Chefs-d'oeuvres de l'éloquence française* (Paris, 1854), 421–26.
41 Louis-Sébastien Mercier, *Eloge de René Descartes* (Paris, 1765), 41–43; Antoine-Léonard Thomas, *Eloge de René Descartes* (Paris, 1765), 4, 67.
42 Voltaire, *Correspondence*, 113: 311 (D12,896); François Azouvi, *Descartes et la France: Histoire d'une passion nationale* (Paris: Fayard, 2002), 124–25.
43 James D. Draper and Guilhem Scherf, *Augustin Pajou: Royal Sculptor, 1730–1809* (New York: The Metropolitan Museum of Art, 1997), 318–20.
44 Béat Louis de Muralt, *Letters Describing the Character and Customs of the English and French Nations* (London, 1726), 8.
45 Brillon de Jouy, *Conclusiones philosophicae* (Paris, 1707); Fritz Saxl, "Veritas Filia Temporis," in *Philosophy and History*, ed. Raymond Klibansky and H. J. Paton (Oxford: Clarendon Press, 1936), 197–222, at 219–21.
46 D'Alembert, *Preliminary Discourse to the Encyclopedia of Diderot*, 89; *Oeuvres de Mr. de Maupertuis*, 4 vols. (Lyon, 1756), 2: 252–53; David Williams, "Condorcet and the English Enlightenment," *British Journal for Eighteenth-century Studies* 16 (1993): 155–69; Jean-Pierre Poirier, *Lavoisier: Chemist, Biologist, Economist*, trans. Rebecca Balinski (Philadelphia: University of Pennsylvania Press, 1993), 333–34.

Chapter 5

1. *Oeuvres de Descartes*, ed. C. Adam and P. Tannery, 12 vols. (Paris, 1964–76), 1: 560, 8A: 3–4; Paula Findlen, "Science as Career in Enlightenment Italy: The Strategies of Laura Bassi," *Isis* 84 (1993): 441–69, at 444. See, in general, Erica Hart, *Cartesian Women: Versions and Subversions of Rational Discourse in the Old Regime* (Ithaca, N.Y.: Cornell University Press, 1992).
2. Wendy Gibson, *Women in Seventeenth-century France* (New York: St. Martin's Press, 1989), 38.
3. François Poulain de la Barre, *Three Cartesian Feminist Treatises*, ed. Marcelle M. Welch, trans. Vivien Bosley (Chicago: University of Chicago Press, 2002), 79–85; Londa Schiebinger, *The Mind Has No Sex? Women in the Origins of Modern Science* (Cambridge, Mass.: Harvard University Press, 1989), 176–77; Siep Stuurman, "Cartesianism: François Poulain de la Barre and the Origins of the Enlightenment," *Journal of the History of Ideas* 58, no. 4 (1997): 617–40.
4. Gibson, *Women in Seventeenth-century France*, 33.
5. Voltaire, *Correspondence*, 85: 104–6 (D92); *The Works of Voltaire: A Contemporary Version*, ed. John Morley, 22 vols. (New York: The Craftsmen of the St. Hubert Guild, 1901), 36: 199.
6. M. de Fontenelle (Bernard Le Bovier), *Oeuvres complètes*, ed. Alain Nidrest, 9 vols. (Paris: Fayard, 1989–2001), 6: 251, 254; David Brewster, *Memoirs of the Life, Writings, and Discoveries of Sir Isaac Newton*, 2 vols. (Edinburgh, 1860), 1: 342; Oliver Goldsmith, *An Inquiry into the Present State of Polite Learning in Europe* [1759], in *Collected Works*, ed. Arthur Friedman (Oxford: Clarendon Press, 1966), 1: 300; Grimm, *Historical and Literary Memoirs and Anecdotes*, 1: 150–51.
7. Marie-Rose de Labriolle, *Le Pour et Contre et son temps*, 2 vols. (Geneva: Institut et Musée Voltaire, 1965), 2: 497; Denis Diderot, "Of Women," in *Dialogues*, trans. Francis Birrell (London: George Routledge, 1927), 196.
8. "He is concerned merely with what women can do, not with what they should do." Pierre Le Moyne, *La galerie des femmes fortes* (Sommavile, 1647), 253, cited in Gibson, *Women in Seventeenth-century France*, 18–19.
9. Nicolas Malebranche, *The Search After Truth*, trans. Thomas M. Lennon and Paul J. Olscamp (Cambridge: Cambridge University Press, 1997), 130.
10. Jean-Jacques Rousseau, *Emile*, trans. Barbara Foxley (London: J. M. Dent, 1993), 418–19.
11. Immanuel Kant, *Observations on the Feeling of the Beautiful and Sublime*, trans. John T. Goldthwait (Berkeley: University of California Press, 1960), 78–79.
12. Nina R. Gelbart, *Feminine and Opposition Journalism in Old Regime France: Le Journal des Dames* (Berkeley: University of California Press, 1987), 42, 57.
13. Pierre-Joseph Boudier de Villemert, *The Friend of Women* [1758], trans. Alexander Morrice (Philadelphia, 1803), 24–27. For more on de Villemert, see David Williams, "The Fate of French Feminism: Boudier de Villemert's *Ami des femmes*," *Eighteenth-century Studies* 14 (1980): 37–55.
14. *Boswell in Holland, 1763–1764, including his correspondence with Belle de Zuylen (Zélide)*, ed. Frederick A. Pottle (New York: McGraw-Hill, 1952), 227.
15. C. Truesdell, "Maria Gaetana Agnesi," *Archive for History of Exact Sciences* 40 (1989): 113–42, at 116–17.
16. *The Works of Mary Wollstonecraft*, ed. Janet Todd and Marilyn Butler, 7 vols. (New York: New York University Press, 1989), 5: 103.
17. Gabriella B. Logan, "The Desire to Contribute: An Eighteenth-century Italian Woman of Science," *American Historical Review* 99 (1994): 785–812; Paula Findlen, "Science as Career in Enlightenment Italy: The Strategies of Laura Bassi," *Isis* 84 (1993): 441–69; Beate Ceranski, *"Und sie fürchtet sich vor niemandem": Die Physikerin Laura Bassi (1711–1778)* (Frankfurt: Campus, 1996); Marta Cavazza, "Laura Bassi 'Maestra' di Spallanzani," in *Il cerchio della vita: Materiali di ricerca del Centro Studi Lazzaro Spallanzani di Scandiano sulla storia della scienza del Settecento*, ed. Walter Bernardi and Paola Manzini (Florence: Leo S. Olschki, 1999), 185–202.
18. Clifford Truesdell, "Maria Gaetana Agnesi,"

Archive for History of Exact Sciences 40 (1989): 113–42; Massimo Mazzotti, "Maria Gaetana Agnesi: Mathematics and the Making of the Catholic Enlightenment," *Isis* 92 (2001): 657–83.

19 Truesdell, "Maria Gaetana Agnesi," 117–18.

20 Paula Findlen, "A Forgotten Newtonian: Women and Science in the Italian Provinces," in *The Sciences in Enlightened Europe*, ed. William Clark, Jan Golinski, and Simon Schaffer (Chicago: University of Chicago Press, 1999), 313–49.

21 Mirella Agorni, *Translating Italy for the Eighteenth Century: Women, Translation and Travel Writing 1739–1797* (Manchester: St Jerome Publishing, 2002), 68; Blandine L. McLaughlin, "Diderot and Women," in *French Women and the Age of Enlightenment*, ed. Samia I. Spencer (Bloomington: Indiana University Press, 1984), 296–308, at 297; Denis Diderot, "Réfutation," in *Oeuvres philosophiques*, ed. Paul Vernière (Paris: Garnier frères, 1961), 606.

22 Charles-Jean-François Hénault, *Mémoires du président Hénault* (repr. Geneva: Slatkine Reprints, 1971), 120; Gabriella B. Logan, "Italian Women in Science from the Renaissance to the Nineteenth Century," Ph.D. thesis (University of Ottawa, 1999), 189.

23 H. J. Mozans, *Women and Science* (Notre Dame, Ind.: University of Notre Dame Press, 1991), 96; Anne-Thérèse de Marguenat de Courcelles, marquise de Lambert, *Oeuvres complètes de madame la marquise de Lambert* (Paris, 1808), 74–75.

24 Mozans, *Women and Science*, 96; *The Complete Letters of Lady Mary Wortley Montagu*, ed. Robert Halsband, 3 vols. (Oxford: Clarendon Press, 1967), 3: 22–23.

25 Theodore Besterman, *Voltaire*, 3rd ed. (Chicago: University of Chicago Press, 1976), 187–88.

26 Emilie du Châtelet, *Institutions de physique* (Paris, 1740), 1–2; Judith P. Zinsser, "Emilie du Châtelet: Genius, Gender, and Intellectual Authority," in *Women Writers and the Early Modern British Political Tradition*, ed. Hilda L. Smith (Cambridge: Cambridge University Press, 1998), 168–90, at 171; Samuel Edwards, *The Divine Mistress* (New York: D. McKay, 1970), 1.

27 Edward Gibbon, *Memoirs of My Life*, ed. Georges Bonnard (New York: Funk & Wagnalls, 1966), 84–85; *The Memoirs of François René Vicomte de Chateaubriand*, trans. Alexander T. de Mattos, 6 vols. (New York: G. P. Putnam's Sons, 1902), 2: 86; Schiebinger, *The Mind Has No Sex?*, 81.

28 *A Woman of Genius: The Intellectual Autobiography of Sor Juana Inés de la Cruz*, trans. Margaret S. Peden (Salisbury, Conn.: Lima Rock Press, 1987), 26; L. L Bucciarelli and N. Dworsky, *Sophie Germain: An Essay in the History of the Theory of Elasticity* (Dordrecht: Reidel, 1980), 10.

29 *Boswell in Holland*, 185; *There Are No Letters Like Yours: The Correspondence of Isabelle de Charrière and Constant d'Hermenches*, trans. Janet Whatley and Malcolm Whatley (Lincoln: University of Nebraska Press, 2000), 48–49, 55, 57, 182, 289, 373, 392; C. P. Courtney, *Isabelle de Charrière (Belle de Zuylen)* (Oxford: Voltaire Foundation, 1993), 53.

30 Cited in George MacDonald Ross, "Leibniz und Sophie Charlotte," in *Sophie Charlotte und ihr Schloß*, ed. S. Herz, C. M. Vogtherr, and F. Windt (Munich: Prestel, 1999), 95–105.

31 Gelbart, *Feminine and Opposition Journalism in Old Regime France*, 108; Bucciarelli and Dworsky, *Sophie Germain*, 13.

32 Mary Terrall, "Gendered Spaces, Gendered Audiences: Inside and Outside the Paris Academy of Sciences," *Configurations* 3 (1995): 207–32.

33 *Yale Edition of Horace Walpole's Correspondence*, ed. W. S. Lewis (New Haven, Conn.: Yale University Press, 1937–), 18: 555; 20: 555.

34 Bathsua Makin, *An Essay to Revive the Antient Education of Gentlewomen* (London, 1673), 3.

35 Molière, *The Learned Ladies*, trans. Renée Waldinger (Great Neck, N.Y.: Barron's Educational Series, 1957). Variations on the theme abound in other literary works: husbands complaining that wives, in their enthusiasm to observe stars and comets, desert the conjugal bed; gamblers complaining that women abandon the card tables to spend their nights with their telescopes; see Gibson, *Women in Seventeenth-century France*, 39.

36 Joseph Addison, *The Spectator*, ed. Donald F.

Bond, 5 vols. (Oxford: Clarendon Press, 1965), 1: 154.

37 Amelia G. Mason, *The Women of the French Salons* (New York, 1891); *Mémoires du président Hénault*, 135.

38 Alphonse Rebière, *Les femmes dans la science: Notes recueillies*. 2nd ed. (Paris, 1897), 354.

Chapter 6

1 Newton, *Principia*, 379–80.

2 *The Poems of Alexander Pope*, ed. John Butt (New Haven, Conn.: Yale University Press, 1963), 808.

3 *The Life, Unpublished Letters, and Philosophical Regimen of Anthony, Earl of Shaftesbury*, 353; Peter Gay, *The Enlightenment: An Interpretation*, 2 vols. (London: Wildwood House, 1973), 2: 26, 1:12.

4 Newton, *Principia*, 380.

5 Georges May, "Observations on an Allegory: The Frontispiece of the *Encyclopédie*," *Diderot Studies* 16 (1973): 159–74, at 164. The composition was reversed in the engraving sent to subscribers.

6 Ibid., 167.

7 Gay, *The Enlightenment*, 2: 132–33.

8 Johann Kunckel (von Löwenstern), *Ars vitraria experimentalis* (Nuremberg, 1743); Martin Warnke, "Works of the Imagination," in *The Scientific Enterprise*, ed. E. Ullmann-Margalit (Dordrecht: Kluwer, 1992), 101–16, at 107, 109.

9 Louis-Sébastien Mercier, *The Night Cap*, 2 vols. (Philadelphia, 1788), 1: 9–10.

10 Monroe Z. Hafter, "Petronius, Mercier, and Goya's Colossus," *Eighteenth-century Studies* 22 (1989): 529–47, at 545 and n. 30.

11 Alfonso E. Pérez Sánchez and Eleanor A. Sayre, *Goya and the Spirit of Enlightenment* (Boston: Bulfinch Press, 1989), 110–16.

12 Paul McCartney, *Henry De La Beche: Observations of an Observer* (Cardiff: National Museum of Wales, 1977), 54.

13 David Williams, "Progress and the Empirical Tradition in Condorcet," *Bulletin de l'Association de Philosophie de Langue Française* 4 (1992): 67–77, at 68.

14 Jean Dhombres and Jean-Bernard Robert, *Joseph Fourier 1768–1830: créateur de la physique-mathematique* (Paris: Belin, 1998), 61.

15 François Arago, *Biographies of Distinguished Scientific Men* (Boston, 1859), 289; Vincent Cronin, *Napoleon Bonaparte: An Intimate Biography* (New York: Morrow, 1972), 153.

16 *Sketches of the History of Man*, 3 vols. (Glasgow, 1819), 1: 102–3, 159; 2: 90.

17 Denis Diderot, *Thoughts on the Interpretation of Nature and Other Philosophical Works*, trans. David Adams (Manchester, England: Clinamen Press, 1999), 37.

18 Peter A. Pav, "Eighteenth-century Optics: The Age of Unenlightenment," Ph.D. thesis (Indiana University, 1964), 54.

19 Craig B. Waff, "Universal Gravitation and the Motion of the Moon's Apogee: The Establishment and Reception of Newton's Inverse-Square Law, 1687–1749," Ph.D. thesis (Johns Hopkins University, 1975), 68–70, 102.

20 Joseph Priestley, *The History and Present State of Electricity, with Original Experiments*, 3rd ed., 3 vols. (London, 1775; repr. New York: Johnson Reprint Corporation, 1966), 1: vii, xv.

21 Franco Venturi, *The End of the Old Regime in Europe, 1776–1789*, 2 vols. (Princeton, N.J.: Princeton University Press, 1991), 1: 418; I. Bernard Cohen, *Revolution in Science*, 514.

22 John F. W. Herschel, *A Preliminary Discourse on the Study of Natural Philosophy*, ed. Andrew Pyle (Chippenham, England: Thoemmes Press, 1996), 274.

23 Jean-Antoine-Nicolas de Caritat, marquis de Condorcet, *Selected Writings*, ed. Keith M. Baker (Indianapolis: The Bobbs-Merrill Company, 1976), 14, 239.

24 *Colloque International et Interdisciplinaire Jean-Henri Lambert* (Paris: Ophrys, 1979); Pierre Speziali, *Physica Genevensis* (Chêne-Bourg: Georg, 1997), 118.

25 *From Natural History to the History of Nature: Readings from Buffon and His Critics*, ed. John Lyon and Phillip R. Sloan (Notre Dame, Ind.: University of Notre Dame Press, 1981), 364; Stephen F. Milliken, "Buffon's Essai d'Arithmétique Morale," in *Essays on Diderot and the Enlightenment in Honor of Otis Fellows* (Geneva: Editions Droz, 1974), 197–206, at 198; Jacques Roger, *Buffon: A Life in Natural History*, trans. Sarah L. Bonnefoi (Ithaca, N.Y.:

Cornell University Press, 1997), 29; Antonio Pereira Poza, "Newton en el pensamiento naturalista de Buffon: la analogia biologica de la Gravedad," *Asclepio* 44 (1992): 207–19.

26 Arthur M. Wilson, *Diderot* (New York: Oxford University Press, 1972), 30, 433; Diderot, *Thoughts on the Interpretation of Nature*, 44, 59, et passim.

27 Maurice Cranston, *Jean-Jacques: The Early Life and Work of Jean-Jacques Rousseau, 1712–1754* (New York: W. W. Norton and Company, 1983), 136–38; *Correspondance complète de J. J. Rousseau*, ed. R. A. Leigh (Geneva: Institut et Musée Voltaire, 1965–95), 1: 73–82; Jean-Jacques Rousseau, *The Social Contract and Discourses*, trans. G.D.H. Cole, J. H. Brumfitt, and John C. Hall (London: J. M. Dent, 1988), 16, 27.

28 Jean-Jacques Rousseau, *Emile*, trans. Barbara Foxley (London: J. M. Dent, 1996), 276, 281–82; Jean-Jacques Rousseau, *Letter to Beaumont, Letters Written from the Mountain, and Related Writings*, ed. Christopher Kelly and Judith R. Bush (Hanover, N.H.: University Press of New England, 2001), 174; John Herman Randall, *The Career of Philosophy*, 2 vols. (New York: Columbia University Press, 1962–65), 253.

29 Newton, *Principia*, 382; Newton, *Opticks*, 405.

30 George Berkeley, *Works*, ed. A. A. Luce and T. E. Jessop, 9 vols. (London: T. Nelson, 1948–57), 7: 225–28.

31 F.E.L. Priestley, "Berkeley and Newtonianism: The *Principles* (1710) and the *Dialogues* (1713)," in *The Practical Vision*, ed. John Campbell and James Doyle (Waterloo, Ont.: Wilfrid Laurier University Press, 1978), 49–70.

32 Pierre Boutin, *Jean-Théophile Desaguliers: un Huguenot, philosophe et juriste, en politique* (Paris: Honoré Champion, 1999), 217–34.

33 Charles de Secondat, baron de Montesquieu, *Oeuvres complètes*, ed. André Masson (Paris: Nagel, 1950–55), 3: 892, 1478; Montesquieu, *The Spirit of the Laws*, trans. Thomas Nugent (New York: Hafner Publishing Company, 1966), 25, 40.

34 *The Writings of Thomas Jefferson*, ed. Andrew A. Lipscomb, vol. 14 (Washington, D.C.: The Thomas Jefferson Memorial Association, 1903), 85–97.

35 Adam Smith, *Essays on Philosophical Subjects*, ed. W.P.D. Wightman and J. C. Bryce (Indianapolis: Liberty Press, 1982), 104–5.

36 Adam Smith, *Lectures on Rhetoric and Belles Lettres* (Indianapolis: Liberty Press, 1985), 145–46; Adam Smith, *An Inquiry into the Nature and Causes of the Wealth of Nations*, ed. R. H. Campbell and A. S. Skinner, 2 vols. (Indianapolis: Liberty Fund, 1981), 1: 75, 77; Norriss S. Hetherington, "Isaac Newton and Adam Smith: Intellectual Links Between Natural Science and Economics," in *Action and Reaction*, ed. Paul Theerman and Adele F. Seeff (Newark: University of Delaware Press, 1993), 277–91.

37 Ferrone, *The Intellectual Roots of the Italian Enlightenment*, 230, 235; Isaac de Pinto, *An Essay on Circulation and Credit*, trans. S. Bagg (London, 1774; repr. Farnborough: Gregg, 1969), 127.

38 Ferrone, *The Intellectual Origins of the Italian Enlightenment*, 256–57; Anthony Pagden, *Spanish Imperialism and the Political Imagination* (New Haven, Conn.: Yale University Press, 1990), 70.

39 Jean Le Rond d'Alembert, *Traité de dynamique* (Paris, 1743); *Oeuvres de Mr. de Maupertuis*, 4 vols. (Lyon, 1756), 42–44.

40 Isabel F. Knight, *The Geometric Spirit: The Abbé de Condillac and the French Enlightenment* (New Haven, Conn.: Yale University Press, 1968), 28, 32–33.

41 Leslie Stephen, *The English Utilitarians*, 3 vols. (London: Duckworth, 1900), 1: 178–79; Charles W. Everett, *The Education of Jeremy Bentham* (New York: Columbia University Press, 1931), 35–36.

42 Edmund Burke, *A Philosophical Enquiry into the Origin of Our Ideas of the Sublime and Beautiful*, ed. J. T. Boulton (London: Routledge and Kegan Paul, 1958), xxviii, 5–6, 11–13, 150; *The Complete Poetry and Prose of William Blake*, ed. David V. Erdman, rev. ed. (Berkeley: University of California Press, 1982), 660.

43 *The Utopian Vision of Charles Fourier*, ed. Jonathan Beecher and Richard Bienvenu (Columbia: University of Missouri Press, 1983), 1; Jonathan Beecher, *Charles Fourier: The Visionary and His World* (Berkeley: University

of California Press, 1986), 65–67.

44 *Mercure de France* (October 1764): 194–95; Christoph Wolff, *Johann Sebastian Bach the Learned Musician* (New York: W. W. Norton, 2000), 6, 9; Elie Halevy, *The Growth of Philosophic Radicalism* (Boston: The Beacon Press, 1955), 9; Henry Higgs, *The Physiocrats* (London: Macmillan, 1897), 101; Luigi Pepe, "Cesare Beccaria et les mathématiques," *Matapli, Bulletin de la Société Mathématique Appliquées et Industrielles* 51 (1997): 47–52, at 47.

Chapter 7

1 Voltaire, *Letters on England*, 69.

2 G. L. Smyth, *The Monuments and Genii of St. Paul's Cathedral, and of Westminster Abbey* (London, 1826), 2: 703–4.

3 Newton, *Correspondence*, 2: 442; Newton, *Principia*, 380.

4 Francis Haskell, "The Apotheosis of Newton in Art," in his *Past and Present in Art and Taste* (New Haven, Conn.: Yale University Press, 1987), 1.

5 James Thomson, *Poetical Works*, ed. J. Logie Robertson (London: Oxford University Press, 1965), 436–42.

6 Richard Glover, "A Poem on Sir Isaac Newton," in Pemberton, *A View of Sir Isaac Newton's Philosophy*, sig. a3–c2.

7 Marjorie H. Nicolson, *Newton Demands the Muse: Newton's Opticks and the Eighteenth Century Poets* (Princeton, N.J.: Princeton University Press, 1946), 38.

8 *Universal Magazine* 3 (1748): 295, cited in Patricia Fara, *Newton: The Making of Genius* (London: Macmillan, 2002), 2; Westfall, *Never at Rest*, 473; Louis-Sébasien Mercier, *Memoirs of the Year Two Thousand and Five Hundred*, 2 vols. (repr. New York: Garland Publishing, 1974), 1: 143–44.

9 Voltaire, *The Age of Louis XIV*, trans. Martyn P. Pollack (London: J. M. Dent, 1969), 375–81.

10 Edward Young, *Conjectures on Original Composition*, ed. Edith J. Morley (Manchester, England: Manchester University Press, 1918), 33–34.

11 William Duff, *An Essay on Original Genius and Its Various Modes of Exertion in Philosophy and the Fine Arts, Particularly in Poetry*, ed. John L. Mahoney (Gainesville, Fla.: Scholars' Facsimiles & Reprints, 1964), vii, 84, 114–15, 119–20, 125.

12 Alexander Gerard, *An Essay on Genius* (repr. New York: Garland Publishing, 1970), 8, 18–19, 35, 44–45, 92–93.

13 Herbert Dieckmann, "Diderot's Conception of Genius," *Journal of the History of Ideas* 2 (1941): 151–82, at 171–74; Claude Adrien Helvétius, *De l'esprit or Essays on the Mind and Its Several Faculties* (repr. New York: Burt Franklin, 1970), 366–67, 420.

14 Immanuel Kant, *Critique of Aesthetic Judgement*, trans. James C. Meredith (Oxford: Clarendon Press, 1911), 169–70.

15 David Hume, *A History of England*, 6 vols. (Indianapolis: Liberty Classics, 1983), 6: 542; Martin Sherlock, *Letters from an English Traveller* (London, 1780), 164.

16 Leonhard Euler, *Letters of Euler to a German Princess*, 2 vols. (repr. Bristol, England: Thoemmes Press, 1997) 1: 82; *The Memoirs of François René Vicomte de Chateaubriand*, trans. Alexander T. de Mattos, 6 vols. (New York: G. P. Putnam's Sons, 1902), 1: 151; Kenneth Dewhurst and Nigel Reeves, *Friedrich Schiller: Medicine, Psychology and Literature* (Berkeley: University of California Press, 1978), 269–70; *The Works of Samuel Johnson*, 11 vols. (repr. New York: AMS Press, 1970), 4: 134.

17 Johann Caspar Lavater, *Physiognomische Fragmente* (Leipzig, 1776; repr. Zurich: Orell Füssli Verlag, 1968), 2: 276–79.

18 Newton, *Correspondence*, 7: 105, 297 n. 2, 165.

19 James P. Muirhead, *The Life of James Watt* (repr. Alburgh: Archival Facsimiles, 1987), 29.

20 Brandon B. Fortune and Deborah J. Warner, *Franklin & His Friends: Portraying the Man of Science in Eighteenth-century America* (Philadelphia: University of Pennsylvania Press, 1999), 58–59, 113.

21 Judith Colton, "Kent's Hermitage for Queen Caroline at Richmond," *Architectura* 2 (1974): 181–91; Cinzia Maria Sicca, "Like a Shallow Cave by Nature Made: William Kent's 'Natural' Architecture at Richmond," *Architectura* 2 (1974): 68–82.

22 René-Louis Girardin, *Promenade ou itinéraire*

des jardins d'Ermenonville (Paris, 1787), 39.

23 Ronald Paulson, *Hogarth*, 3 vols. (New Brunswick, N.J.: Rutgers University Press, 1992), 2: 2–3.

24 *The Papers of Benjamin Franklin*, ed. Leonard W. Labaree et al. (New Haven, Conn.: Yale University Press, 1959–), 3: 348–49; Fortune and Warner, *Franklin & His Friends*, 26–29.

25 Brooke Hindle, "Cadwallader Colden's Extension of the Newtonian Principles," *William and Mary Quarterly*, 3rd ser., 13 (1956): 459–75; Fortune and Warner, *Franklin & His Friends*, 37–41.

26 *The Literary Diary of Ezra Stiles, D.D., LL.D.*, ed. Franklin B. Dexter, 3 vols. (New York: Charles Scribner's Sons, 1901), 1: 131–32; Fortune and Warner, *Franklin & His Friends*, 32–34.

27 Fritz Wagner, *Zur Apotheose Newtons Künstlerische Utopie und naturwissenschaftliches Weltbild in 18. Jahrhundert* (Munich: Bayerischen Akademie der Wissenschaften, 1974), 13–15; Josef Strasser, *Januarius Zick 1730–1797: Gemälde-Graphik-Fresken* (Weißenhorn: Anton H. Konrad Verlag, 1994), 28.

28 James Barry, *An Account of a Series of Pictures, in the Great Room of the Society of Arts, Manufactures, and Commerce* (London, 1783), 360–63.

29 Manuel, *A Portrait of Isaac Newton*, 193; Domson, *Nicolas Fatio de Duillier and the Prophets of London*, 33.

30 Helen Rossenau, *Boullée & Visionary Architecture* (New York: Harmony Books, 1976), 107–8; Richard A. Etlin, *The Architecture of Death: The Transformation of the Cemetery in Eighteenth-century Paris* (Cambridge, Mass.: MIT Press, 1984), 130–36.

31 Etlin, *The Architecture of Death*, 141–44.

32 Philippe Bordes, "Un élève républicain, Robert-Guillaume Dardel," in *Augustin Pajou et ses contemporains* (Paris: Musée de Louvre, 1999), 509–36, at 520; Bent Sørensen, "The Apotheosis of Sir Isaac Newton by Piranesi," *Apollo* (2001): 26–34.

33 Johann Wolfgang von Goethe, *Conversations with Eckermann* (New York: M. Walter Dunne), 299.

SUGGESTED READING

Buchwald, Jed Z., and I. Bernard Cohen, eds. *Isaac Newton's Natural Philosophy.* Cambridge, Mass.: MIT Press, 2001.

Cohen, I. Bernard. *The Cambridge Companion to Newton.* New York: Cambridge University Press, 2002.

Cohen, I. Bernard. *Newton: Texts, Backgrounds, Commentaries.* Edited by Richard S. Westfall. New York: W. W. Norton & Company, 1995.

Dobbs, Betty Jo Teeter. *The Janus Faces of Genius: The Role of Alchemy in Newton's Thought.* Cambridge: Cambridge University Press, 1991.

Fara, Patricia. *Newton: The Making of Genius.* London: Macmillan, 2002.

Fauvel, John, et al., eds. *Let Newton Be!* New York: Oxford University Press, 1988.

Ferrone, Vincenzo. *The Intellectual Roots of the Italian Enlightenment: Newtonian Science, Religion, and Politics in the Early Eighteenth Century.* Atlantic Highlands, N.J.: Humanities Press, 1995.

Guicciardini, Niccolò. *Reading the Principia: The Debate on Newton's Mathematical Methods for Natural Philosophy from 1687 to 1736.* Cambridge: Cambridge University Press, 1999.

Hall, A. Rupert. *Philosophers at War: The Quarrel Between Newton and Leibniz.* Cambridge: Cambridge University Press, 1980.

Heilbron, John. *Electricity in the 17th and 18th Centuries: A Study of Early Modern Physics.* Berkeley: University of California Press, 1979.

Israel, Jonathan. *Radical Enlightenment: Philosophy and the Making of Modernity, 1650–1750.* New York: Oxford University Press, 2002.

Manuel, Frank Edward. *Isaac Newton, Historian.* Cambridge, Mass.: Harvard University Press, 1963.

Manuel, Frank Edward. *A Portrait of Isaac Newton.* Cambridge, Mass.: Harvard University Press, 1968.

Manuel, Frank Edward. *The Religion of Isaac Newton.* Oxford: Clarendon Press, 1974.

Newton, Isaac. *Opticks: or, A Treatise of the Reflections, Refractions, Inflections, and Colours of Light.* Mineola, N.Y.: Dover Publications, 1952.

Newton, Isaac. *The Principia: Mathematical Principles of Natural Philosophy.* Trans. I. Bernard Cohen and Anne Whitman. Berkeley: University of California Press, 1999.

Porter, Roy. *Creation of the Modern World: The Untold Story of the British Enlightenment.* New York: W. W. Norton & Company, 2001.

Westfall, Richard S. *Never at Rest: A Biography of Isaac Newton.* Cambridge: Cambridge University Press, 1980.

ACKNOWLEDGMENTS

It is a duty and a pleasure to enumerate the many debts I have incurred in writing this book. Roger Ariew, Jed Buchwald, and George Smith were constant companions in the process of composition as well as careful readers of several chapters. Many other friends and colleagues generously assisted in my research, responded to queries, and helped in procuring images: Brian Allen, Ugo Baldini, Jonathan Beecher, Elena Brambilla, Ron Calinger, Miguel Carolino, Ginevra Carsignani, Paolo Casini, Marta Cavazza, Edward Conant, Robert Essick, James Evans, Cinzia Ferrini, Vincenzo Ferrone, Paula Findlen, Peter Fox, Charles Gillispie, Robert Gordon, Anthony Grafton, Susan Green, Rob Iliffe, Jack Iverson, Lisa Jardine, Milo Keynes, Kevin Knox, Bettina Koch, Janis Langins, Henrique Leitao, Rhodri Lewis, Scott Mandelbrote, Massimo Mazzotti, Renato Pasta, Ronald Paulson, David Ruderman, J. B. Shank, Paul Shore, Stephen Snobelen, Mariafranca Spallanzani, Barbara Stafford, Anthony Turner, Reinhard Uhrig, Rienk Vermij, Françoise Waquet, Janet Whatley, and Judith Zinsser. Special thanks is owing to Shady Peyvan and her staff at the Interlibrary Loan system at Caltech who managed to get for me every book and article I required; I would not have been able to carry out my research without them.

I would like to express my sincerest gratitude to librarians and curators at several institutions who not only provided me with the illustrations adorning this volume, but did their utmost to waive fees, or at least keep them reasonable. In particular, I wish to acknowledge the kindness of Adam Perkins of the Cambridge University Library and Ben Weiss of the Burndy Library, who devoted many hours to surveying material in their possession, and patiently responded to my many requests. I am also indebted to Judith Goodstein and Shelley Erwin at the California Institute of Technology Archives; Bruce Whiteman at the William Andrews Clark Library and Marina Romani at the UCLA Center for 17th- and 18th-century Studies; Bruce Stephenson and Devon Pyle-Vowles at the Adler Planetarium; Dennis Marnon and Hope Mayo at the Houghton Library; Sara Schechner at the Collection of Historical Scientific Instruments at Harvard University; Ron Brashear at the Smithsonian Institution; Jenae Huber at the Tacoma Art Museum; Laura Barry at the Colonial Williamsburg Foundation; Anna Louise Ashby at the Pierpoint Morgan Library; Cynthia Roman at the Lewis Walpole Library; Harmony Haskins at the White House Historical Association; Eliana Moreira at the Bridgeman Art Library; David McKitterick and Jonathan Smith at Trinity College Library, Cambridge; Christine Woolett at the Royal Society Library; Emma Butterfield at the National Portrait Gallery, London; Robin Hamlyn and Katie Dobson at the Tate Gallery; Tom Sharpe at the National Museum of Wales; Ian Milne at the Royal College of Physicians, Edinburgh; Alice Nørhede at the Danish Pharmaceutical Library; Luisa Pigatto at the Museo *La Specola*, Padua; Libuse Piherova at the National Library of the Czech Republic; and Henri de Breteuil.

Special thanks are owed to the friends I made in the Publications Office of The New York Public Library – Barbara Bergeron, Anne

Skillion, and Karen Van Westering – who not only shared my enthusiasm for the subject matter but waited, with considerable restraint, for the arrival of the manuscript. Ken Benson, Marc Blaustein, Kara Van Woerden, Suzanne Doig, and Jennifer Woolf contributed a great deal to the design and production of the book.

The book was conceived as a companion volume to an exhibition at The New York Public Library, and I would like to thank Paul LeClerc and H. George Fletcher for inviting me to curate the exhibition and for encouraging me to be ambitious. Jeanne Bornstein and Meg Maher furnished me with a jump-seat in their tiny office and gently guided me through the fine art of curating. Susan Rabbiner weathered fiscal and other tribulations with a wry sense of humor, while the cheerful members of the Registrar's office – Jean Mihich, Caryn Gedell, Jessica Glasscock, Julie Joseph, and Lori Mahaney – never demurred at the constantly shifting contours of the exhibition. For their part, Myriam de Arteni and Russell Drisch and their staffs – Shane Caffrey, Patrick T. Day, Susan Fisher, Andrew Gaylard, Margaret Greene, Eric Kidhardt, Rachel McPherson, Ana Mari de Quesada, Sarah Seigel, Ben Shambaugh, Peter Wagner, and Tom Zimmerman – expertly handled conservation, preparation, and display. Last, but certainly not least, Stephen Saitas conceived an exquisite design for the exhibition that wonderfully captured the spirit of the Newtonian moment.

As always, my greatest debt goes to my family, which lived with "Sir Isick," in various incarnations, for a long time. Carol endured a great deal, and without her encouragement and assistance the book would not have been ready. Equally supportive were Koby and Ariella who, in addition, contributed much needed merriment to the "history of the fork."

The book is dedicated to Frank Manuel and Bernard Cohen, two superb historians who, in very different ways, made some of the most significant contributions to our understanding of Newton's life, ideas, and legacy during the twentieth century, and whose companionship I sorely miss.

CREDITS

The following divisions and collections of The New York Public Library (NYPL)'s Research Libraries are represented in this volume:

HUMANITIES AND SOCIAL SCIENCES LIBRARY
Art & Architecture Collection, Miriam and Ira D. Wallach Division of Art, Prints and Photographs
Henry W. and Albert A. Berg Collection of English and American Literature
General Research Division
Carl H. Pforzheimer Collection of Shelley and His Circle
Print Collection, Miriam and Ira D. Wallach Division of Art, Prints and Photographs
Rare Books Division
Spencer Collection

THE NEW YORK PUBLIC LIBRARY FOR THE PERFORMING ARTS
Music Division

SCIENCE, INDUSTRY AND BUSINESS LIBRARY

CREDITS

pp. 9, 15, 49, 63 (nos. 3a-b), 68, 69, 71, 74, 82 (nos. 7–8), 83 (no. 11), 84, 87 (right), 92 (bottom), 98, 103, 109, 110, 128, 132, 134, 154, 158, 160, 161, 179: The Burndy Library, Dibner Institute for the History of Science and Technology, Cambridge, Massachusetts
p. 28: Photograph by Jeremy Whitaker
p. 62 (right): Typ 820.10.7582, Department of Printing and Graphic Arts, Houghton Library, Harvard College Library
p. 63 (top right): The Metropolitan Museum of Art, Thomas J. Watson Library, Gift of Jane E. Andrews in memory of her husband, William Loring Andrews. Photograph © 2004 The Metropolitan Museum of Art
p. 80: *IC7.V6643.725pc, Department of Rare Books, Houghton Library, Harvard College Library
p. 114: The Metropolitan Museum of Art, Gift of Georgiana W. Sargent, in memory of John Osborne Sargent, 1924. [24.63.1116(17)] Photograph © 2004 The Metropolitan Museum of Art
p. 116: The Metropolitan Museum of Art, Purchase, Mr. and Mrs. Charles Wrightsman Gift, in honor of Everett Fahy, 1977. (1977.10) Photograph © 1989 The Metropolitan Museum of Art
p. 136 (left): Typ 720.61.804(A), Department of Printing and Graphic Arts, Houghton Library, Harvard College Library
p. 165 (no. 4): The Pierpont Morgan Library, New York. Peel 0401. Photography: Joseph Zehavi, 2004.
p. 165 (no. 6): EB7 W5794 F825m, Department of Rare Books, Houghton Library, Harvard College Library
p. 174 (bottom left): Photograph by Jeremy Whitaker
p. 180 (top): The Pierpont Morgan Library, New York. Peel 0412. Photography: Joseph Zehavi, 2004
p. 182: Yale University Art Gallery, Bequest of Dr. Charles Jenkins Foote, B.A. 1883, M.D. 1890
p. 189: The Metropolitan Museum of Art, The Elisha Whittelsey Collection, The Elisha Whittelsey Fund, 1951. (51.501.2843) Photograph, all rights reserved, The Metropolitan Museum of Art

INDEX